TEMPO

http://tempobook.com

#tempobook

TEMPO

Timing, tactics and strategy in
narrative-driven decision making

Venkatesh Rao

Ribbonfarm, Inc.

First Edition

© 2011 by Ribbonfarm, Inc. All rights reserved

ISBN: 978-0-9827030-0-7

http://ribbonfarm.com

Cover design and artwork by Adam Hogan

First Edition

Printed and distributed by Lightning Source

For Wilbur, a high-tempo cat.
2007 - 2008.

Lonely one, you are going the way to yourself. And your way leads past yourself and your seven devils. You will be a heretic to yourself and a witch and soothsayer and fool and doubter and unholy one and a villain. You must wish to consume yourself in your own flame: how could you wish to become new unless you had first become ashes!

Lonely one, you are going the way of the creator: you would create a god for yourself out of your seven devils.

– Fredrick Nietzsche, Thus Spake Zarathustra

Contents

Preface

Cooking was my unwinding ritual during the many leisurely years I spent studying the decision sciences, first as a graduate student at the University of Michigan and then as a postdoctoral researcher at Cornell. This book originally started – and I am being completely serious here – as an idea for a cookbook. Here's the story of how it came to be a book about decision-making.

My time at Cornell, between 2004 and 2006, was ridiculously idyllic. For a good part of that period, I lived a rather monastic life in an isolated apartment complex near the Buttermilk Falls state park, in the hills outside Ithaca. Not quite Thoreau's Walden Pond, but close. Every evening, I'd drive back from my office on campus past several of Ithaca's famous waterfalls. Whether the day had been frustrating or rewarding, the sight of the waterfalls would always return me to equilibrium. Soon after I settled into this routine, I noticed that dinners started in this state of waterfall-induced equilibrium seemed to turn out better. And so it was that my times in the kitchen turned into episodes of idle wondering, as I chopped and stirred, about the role of waterfalls and other subtle ingredients in cooking.

My memories of Ithaca revolve around these solitary episodes in the kitchen. I had plenty of time to ponder the broad question of how cooks make decisions in kitchens, ranging from *what can I make with what's in my fridge?* to *is it time to flip this pancake?* Out of these episodes was born an idea for a book on improvisational cooking, and the rhythms and cadences of the kitchen.

But as I gathered my thoughts and began making my notes, I realized that as a cook, I was at best an enthusiastic amateur. What I really wanted to do, it became clear, was write about the interplay of rhythms, emotions and energy in decision-making that is so vividly evident in cooking. The kitchen was just a particularly rich metaphor for a broader

1

theme.

So I started assembling the ideas into a broader synthesis, and began noticing and collecting questions, ideas and examples from other domains. It seemed to me that the rich interplay of rhythms, emotions and energy was everywhere I looked:

1. How was I making the apparently trivial decision of when to flip a pancake, against the backdrop of impatience, hunger, wrist-skill, and flame control?

2. How, by focusing on my stroke and breathing, rather than my strength, was my swim coach able to lower my sprint times *and* the amount of energy I spent?

3. How do musicians performing jam sessions decide what to play next?

4. How do good managers steer a meeting?

This was a complex set of questions, and I found myself groping for that key insight, that high-concept which would miraculously dissolve the complexity. The answer began to emerge from the research I was involved in at the time, on command and control.

My research, funded by the US Air Force, concerned what are called mixed-initiative command and control models: complex systems where humans, autonomous robotic combat vehicles and software systems share decision-making authority.

Somewhere in my reading, I stumbled across the idea that one must control the tempo of a military engagement. Shock-and-awe is an example of tempo-driven war-making.

The idea, and in particular, the word *tempo*, stuck in my mind as clever and evocative. But at the time, I didn't pay much attention; it seemed like no more than a clever motif. It would take me nearly four more years, including two immersed in the decidedly non-monastic world of business, before I recognized the idea of tempo for what it was: the high concept I was looking for, to organize my messy thoughts on decision-making that began in the kitchen.

I first attempted a synthesis towards the end of my postdoctoral stint with Raff D'Andrea's group at Cornell. In 2006, I designed and taught a graduate course on the decision sciences, in collaboration with my doctoral advisor Pierre Kabamba, who taught the course in parallel at the University of Michigan. Among the workshop exercises I devised for the course were a challenge to model call-and-response drumming, and at least one example involving the kitchen. It was a lot of fun.

Yet, despite the fun, the course had a discursive, grab-bag feel to it. One of the smartest students in the class (and a key architect of the world-beating Cornell robotic soccer team that Raff managed at the time) commented, "I like the material, but I don't yet see the thin red line connecting all the ideas."

He was right. I was missing certain powerful ideas about metaphor and narrative in human decision-making that had been evolving in the humanities since the 1970s. When I left Cornell and joined Xerox in 2006, I had an opportunity to dive into those subjects. At Xerox, I encountered the fascinating world of collective, real-time business decision-making. At the same time, I started my blog, ribbonfarm, which allowed me to tackle the ideas from the humanities that I'd encountered, but not explored in depth, during my university years.

And one fine day, in the summer of 2008, I realized that *Tempo* was the title of a book demanding to be written. A book that would weave the tragic passion of the humanities together with the unsentimental austerity of the mathematical decision sciences, to tell you the story of your life.

The ideas in this book were inspired by questions from everyday life at work, home and play. I have tried to return to those worlds, and frame what I have to say as deeper insights into common sense. But this has not always been possible. Ideas from dusty academic journals and from rarefied domains of practice – such as boardrooms and the decks of aircraft carriers – are not always reducible to everyday experience. So yes, at times, you will need to stretch your mind.

But you should expect and even welcome the challenge. You live in a complicated world, where you must contend with globe-spanning human decision processes that affect your life. Everything you need to know, you did not learn in kindergarten.

Welcome to Tempo.

Chapter 1

Introduction

Does the road wind up-hill all the way?
Yes, to the very end.
Will the day's journey take the whole long day?
From morn to night, my friend.

From *Up-Hill* by Christina Rossetti

Decisions punctuate our thoughts and actions and set the pace of our lives. They drive, and are in turn driven by, energy and emotional experience – the ebb and flow of anxiety, lethargy, excitement and impatience – and lend to our experience of time a choppy, emotionally charged *tempo*. This notion of tempo is the central concept we will use to organize this book about decision-making. There are many excellent reasons to look at decision-making from this perspective, but perhaps the best reason is that it is somewhat unsettling. My objective in this book is to get you to look at the familiar with new eyes.

Watch a modern information worker through the glass the next time you walk past a coffee shop. All you can see is typing, mouse-wielding and cell-phone use. Information work blurs, and sometimes eliminates, the distinction between thought and action, and elevates the role of decision-making above both. Frederick Brooks, in his 1975 classic, *The Mythical Man Month*,[1] noted that "the programmer, like the poet, works only slightly removed from pure thought-stuff." Today every sort of information work is acquiring the pure thought-like characteristics Brooks first noticed in programming. In these conditions, separating work into thought and action is less useful than it used to be. Work is simply whatever we must do to get from one decision to the next.

Momentum accumulates and degrades constantly during the dozens of micro-decisions we make every minute. The world changes so fast that most of these decisions go towards simply maintaining situation awareness: largely subconscious decisions about what beliefs to add, update or discard, a process that computer scientists call *truth maintenance*.

For those decisions that *do* affect visible behavior, the classic metaphor of a fork in the road is inadequate. A better metaphor is the sort of continuous and complex decision-making involved in driving: as Tom Vanderbilt notes in *Traffic*,[2] at 30 miles per hour, we are exposed to about 1,320 "items of information" per minute.

You cannot manage this process one decision at a time. Neither can you selectively override your subconscious to only make the big decisions. This is the main reason tempo is a useful notion: it varies slowly enough that you can maintain an awareness of it, and use it to modulate the blurry torrent of your life. Being aware of tempo allows you to manage the momentum of your life. Managing momentum also means that when you *do* choose to slow down and pluck out individual consequential decisions for conscious and detailed processing, you won't be the victim of conjuring tricks pulled on you by the mob of unruly autopilots – what Marvin Minsky called the "Society of Mind" – that exists just below conscious awareness.[3] The more you consciously manage tempo, the more creative and realistic your options during the big decisions.

Information work also elevates the roles of *when, where* and *who* above the roles of *what, why* and *how* in decision-making. Most of what has been written about decision-making, both popular and scholarly, has focused on the latter triad. *What* gives you the study of options. *Why* gives you the study of causation, motivation, reward and punishment. *How* leads you to the classic problems of means-ends reasoning, such as planning and scheduling.

For much of history, *when, where* and *who* – questions about timing, framing, background and context – have been considered trivial. Pre-modern life was driven by the rhythms and seasons of the natural world, and contained within local geographies, mostly static organizational structures and social networks. Answers to *when, where* and *who* have usually been along the lines of "in due season," "down the street" and "at the village council."

Contrast this with the act of buying a book on Amazon.com, an iconic instance of modern decision-making: the site's recommendation and review systems can strongly influence what you buy and why, but *when*, *where* (a shipping address is a non-trivial piece of data for modern nomads) and *for whom*, (for yourself, or for someone in your global social network) are decisions you control more strongly than your ancestors.

Today, as virtualization mangles our intuitive notions of time, space and social context, such gradual and subtle shifts are evident everywhere. Many of the surprising insights in Bill Tancer's *Click*,[4] a book about what web traffic analysts see people doing online, have to do with these *when, where* and *who* questions. Among the six *W* questions (*What, Why, hoW, When, Where* and *Who*) that are involved in any decision, *when* today stands supreme as the fundamental one. If you restrict yourself to just the first three questions, instead of managing all six, you will find yourself in the world of the Red Queen in *Alice in Wonderland*. You will need to run faster and faster, just to stay in the same place. If you learn to consciously manage tempo, however, you will be able to achieve the miracle of staying ahead of the fast-paced world, while slowing down.

Though *when* is the driving question in this book, *where*, appropriately generalized for our post-geographic virtualized world, will play a major supporting role. The other four *W* questions will appear as well, though in forms that may be unfamiliar to you.

Within this modern context, we will talk about all sorts of decisions. Here is a sampler of what is to come:

1. Should I respond to this email now or later?

2. Should I sleep in and blow off the 8:00 AM classs?

3. Should I schedule a special meeting or wait for the routine weekly meeting?

You will notice a bias towards micro-decisions. For bigger, more consequential decisions, we will adopt a framework based on synthesis, design, metaphor and storytelling, rather than selection among predefined options. So rather than treating college as a "which college should I attend, and what major should I pick?" decision, we will treat it as the creative process of continuously retelling and enacting the most compelling *College* story you can. The larger the scale, the more the six *W* questions will recede into the background.

We will approach our subject from several useful angles, from the commonsensical to the scientific to the philosophical. We will explore the idea of mental models and drill down into the notions of planning and procedure. We will talk about the semantics of the word *strategy* and tease out the substance underlying this seemingly vacuous term in everyday use. We will probe the subtleties of risk and uncertainty.

But as we navigate through this complexity, the idea of tempo will remain our true north. Again and again, we will return to this idea and use it anchor our understanding of other concepts. The idea of tempo will help us reduce theoretical concepts to practical principles, and help codify the lessons of specific anecdotes in general ways.

We will start slow, since this perspective is an unfamiliar one. But once we are warmed up, I will set a demanding (and occasionally, punishing) pace through some rough and beautiful uphill territory. But if you can keep up, the view at the top will be worthwhile, I promise you.

1.1 Tempo Characterized

I define tempo as the set of characteristic rhythms of decision-making in the subjective life of an individual or organization, colored by associated patterns of emotion and energy.

Throughout this book, I will draw on everyday domains, such as the kitchen and the workplace, to illustrate concepts and principles. Occasionally, for variety, I will draw on other domains, including time-tested ones such as war and sports, and less familiar ones like music and traffic.

Let us start with a few example domains.

1.1.1 Tempo in the Kitchen

The kitchen is an excellent decision-making laboratory, since it is a real-world domain, but small enough to be tractable. It is commonplace and familiar to most of us. It illustrates every important idea we will talk about, and works both as an example and as a metaphor for other domains. Our experience of the kitchen (unlike, say, our experience of politics or global warming) is relatively complete and self-contained. In the kitchen, we personally encounter everything from ideation, through execution, to consequences intended and unintended. Let's start by looking at an aspect of tempo in the kitchen.

> A sleepy restaurant in a small town is faced with an unexpected onslaught of diners from a tour bus while the executive chef is out running an errand. The staff starts to panic, and lose its collective head. Ten minutes into the confusion, the executive chef returns, and instantly gets that there is a "situation." He barks orders: "You! get the water boiling for the pasta. You there, man the vegetable station and get started chopping the tomatoes. We're almost out of potatoes. You, run down to the store and get another 10 lb bag."

This is an example of a pattern of tactical decision-making that we'll call *scan-to-task*.* You look around and rapidly assign every open resource you see to an open problem. Gordon Ramsey, the celebrity chef, demonstrates this pattern very effectively on his television shows.

There is a lot going on here. Several phenomena associated with decision-making, such as situation awareness, anticipatory planning and resource management are evident. But at the heart of the evolution of the situation is a change in the group's collective tempo at the eleventh minute, when the executive chef turns confusion and anxiety into action. The psychological clock ticks faster, attempting to catch up with the real world. If you were to experience the episode and tell the story later, this would be a key moment in your narrative. This is not an accident. Shifts in tempo are central to the model of rationality I will develop in Chapter 4: narrative rationality.

1.1.2 The Workplace

Work is a tougher domain to analyze than the kitchen. Much is hidden within the subtleties of language. Execution is a distant abstraction (often happening as far away as factories on the other side of the planet). Unintended consequences are hidden within the obscurity of international carbon-credit trading schemes rather than visibly accumulating in the kitchen trash can.

But tempo changes are also central to the world of work, as we will see. Here is a simple example.

> A highly charged and stimulating business meeting is in progress. There is a sense of urgency. Agenda items are being raised,

---

*A friend, a navy veteran, introduced me to this phrase, while describing the leadership style of one of his COs.

creative options are being quickly proposed, and decisions are being rapidly made. Quietly at first, and then more forcefully, a hitherto silent participant interrupts to point out a tricky ethical issue that is being glibly ignored. Like dominoes falling, the participants shut up one by one and turn their attention to the interrupter. The symphony of effective deliberation and decision-making dies down, first to an uncertain murmur, and then to silence.

Again, there is a lot going on. Assumed consensus, reframing, the interplay of values and decisions – are all in evidence. Note that a change in tempo marks a critical turning point in the deliberations. The interruption and its aftermath will likely be remembered as the highlight of the meeting.

1.1.3 Personal Life

Let's round off this set of examples by moving over to personal life. Business books rarely pay much attention to this domain, but modern information workers (free agents and virtual workers in particular) blend work and life in such intimate ways that it is impossible to manage the tempo of either in isolation. Our example is the sort of clichéd episode that might appear in a romance novel.

An experienced Casanova is on a date with an equally sophisticated woman. Artfully he manages the pace of the evening, plying his companion with wine. His conversation is masterful; he maneuvers from comforting, relaxing topics to piquant ones. Our Casanova acknowledges the theatricality of the evening with gentle, ironic humor, but he is careful not to let irony overwhelm suspension of disbelief. At exactly the right moment, he sweeps her to the dance floor. The evening winds down, and they walk out of the restaurant, laughing, his arm around her waist, to the parking lot. A moment of silence follows as the laughter trails off and they approach the car. It is just slightly uncomfortable and physically awkward as he must let go of her waist to open the door for her. That's the opportunity he's been trying to casually engineer through the evening. With a hint of dis-

arming fumbling, he leans in for a kiss, just stopping short,
to let her make the decision.

Clichéd though it is (though you've probably met men and women
who can actually dance this dance), the hypothetical vignette is illumi-
nating. The entire episode is characterized by the control (and letting
go of control) of tempo and momentum, across a broad spectrum from
relaxed to urgent.

While the man nominally leads, the narrative of the evening is being
co-created and managed by both the man and the woman. Through her
deceptively lower-energy, reactive role, the woman exercises as much
control over the proceedings as the man does through his more active role. Overt decision-making is almost absent. The only meaningful
fork-in-the-road moment is the kiss-decision at the end, but the rest of
evening has been so expertly managed, the outcome is irrelevant. The
narrative can end gracefully whatever the woman chooses to do.

This represents a pinnacle of artistry where your mind is in what I
like to call the *clockless clock* state.[†] Keep this phrase in mind because
we will glimpse more of the clockless clock in domains such as stand-up
comedy and improvisational theater. Ultimately, all the concepts, theories, ideas, exercises and examples in the book are merely scaffolding.
This state is what we really want to explore.

1.2   Skill: Tempo Doodling

In every chapter of this book we will discuss a variety of behaviors,
skilled and unskilled. Each chapter will explain one key behavior at a
level of detail that will allow you to actually practice and acquire it.
Many of these skills are based on fairly sophisticated ideas, and require
learning some concepts first. But the one you've just encountered can
be understood and practiced without much theory: detecting changes in
tempo in your environment and becoming sensitive to how they affect
your own emotions and energy.

Wherever you are right now, be it at your desk, on a couch watching
TV while reading, on an airplane, or in a café, see if you can detect the

---

[†]Readers familiar with the elements of Zen will no doubt recognize a kinship with the
duality-busting, self-destructing nature of its concepts. This phrase was inspired by the
collection of Zen koans, *The Gateless Gate*.

next change in tempo. Some of these can be subtle, like your refrigerator compressor turning itself on, or the switch from a subdued moment in your favorite sitcom to an excited commercial. Was there some restless shifting on the plane when the flight attendants brought out the beverage cart? Did that listless gang of young men at the table next to you in the café perk up when the pretty young woman walked in?

You just observed tempo in the raw, along with all sorts of other distracting features. To develop a skill, you must first isolate it, just as you use dumbbells to isolate and work specific muscles. Here are your dumbbells for tempo detection, a doodle alphabet inspired by the fascinating art of Amy Lin,[‡] whose colored-pencil pieces are composed entirely of dots.

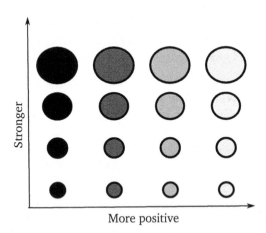

In our tempo alphabet, empty circles are positive-emotion events, shaded circles are negative-emotion events. Bigger is more energetic. Closer together is faster. To practice this skill, doodle in cafés and meetings. Here is an example phrase.

[‡]http://amylinart.com

And here is a sketch of an actual situation, such as one person talking to another, and steadily getting angrier. You might scribble something like this while watching an irate customer at a customer service desk, with one symbol per conversational turn.

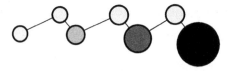

Can you tell which dots represent the customer? Don't worry about being precise. If the only way you can represent growing excitement is to have symbols overlap, let that happen. Try something more subtle, like a boring conversation that starts politely, but eventually becomes tired and slow-paced.

As you play your private game of watching tempo changes in your favorite environments, pay attention to how your own emotions and energy levels respond. If you are up to watching two things at once, try capturing the tempo inside your head as well. Come up with your own doodle alphabet or use colored pens if you like. Aim for art, not precision, and save particularly cool doodles. You can give them evocative names like "Angry Dad yelling at his daughter."

Ultimately, this exercise is just a warm-up. Beyond a point, you will not need to doodle; you can respond with your body. One exercise I find useful is to hold on to different levels of tension in your body, from complete relaxation to spikes of frozen alertness. By consciously holding on to your bodily reactions to environmental tempo changes, for a little longer than they would normally last, you can train yourself to be more aware of them. This turns your entire body into a tempo barometer.

As you try such exercises, you will find that becoming aware of internal and external tempos calms you, thereby muting your reactions. An instrument that systematically under-reacts is no good, so you can try a more challenging exercise: deliberately over-reacting to tempo changes. Practice being jumpy and on-edge for ten minutes, just to experience the feeling. It can be quite enlightening.

To improve your skills, it helps to occasionally look through more sensitive eyes. That means consuming art. Rent and watch Walter Ruttmann's 1923 classic black-and-white movie, *Berlin: The Symphony of a Great City* (*Berlin: Die Symphonie der Großstadt*). To reinforce this skill, I will make more art suggestions throughout the book.

Learning tempo awareness is easy. The next behavior will require more work.

1.3   A Note on Military Thought

A note is in order, to readers familiar with the historical development of military thought. Those who lack this familiarity can safely skip this section.

Many of the ideas in this book were inspired by my work at Cornell, sponsored by the US Air Force, between 2004 and 2006. These range from the use of tempo as an overarching organizing concept, to specific ideas such as my definition of situation awareness, which is directly borrowed from the military literature. At a broader level, the classic ideas of Clausewitz,[5] Mahan[6] and Boyd[7] have been instrumental in shaping the decision-making models in this book, and this influence will be evident to knowledgeable readers.

Besides this formal influence, many of the informal ideas and anecdotes that shaped my thinking originated in conversations with military veterans.

Despite this influence, I have chosen to develop the material primarily using everyday civilian examples, for three reasons.

The first is that I want to keep the book accessible to those without specialized knowledge of military matters. As I discovered through feedback on an early draft, even the simplest ideas from domains like air combat tend to confuse and distract the general reader.

The second reason is that I prefer to tread with caution in a domain where I myself am at best an enthusiastic amateur, working with somewhat obsolete information. Too little time has passed for the lessons of early twenty-first century military conflicts to be codified into revised doctrines, but it is already clear that revisions will in fact be needed. It seems unwise to rely too heavily on material in obvious need of significant updates.

And finally, I wish to avoid diving too deeply into the specific characteristics of adversarial situations, which are the primary concern of military thought. While there is an adversarial element in most socially situated decision-making, usually that element does not require much special handling.

That said, this book originated in conversations with thoughtful, creative and reflective members of the military, and I look forward to learning from any further such conversations it might spark.

Chapter 2

A Sense of Timing

This then, was the Great Invention; the use of oscillatory motion to track the flow of time. One would have expected something very different – that time, which is itself continuous, even and unidirectional would be best measured by some other continuous, even and unidirectional phenomenon.

- David Landes, *Revolution in Time*

Ask someone who has just changed jobs, or traveled to a different country, about their impressions. If a change of tempo was involved, chances are that will be the first thing they mention. "Things are much more relaxed there," or "it is a very exciting and fast-paced lifestyle." In fact the decision to switch jobs or pick a particular vacation destination was likely motivated by the allure of a different tempo.* "I wanted a change of pace," your friend might say. Some professions are even commonly defined in terms of their characteristic tempo. Pilots, policemen and firefighters describe their work as "long periods of sheer boredom punctuated by moments of sheer panic."

Tempo has three elements: rhythm, emotion and energy. Western musical concepts surrounding tempo – *adagio, andante, allegro* and *presto* – convey more than mere pace. They also indicate mood and energy.

Even the rhythm component of tempo is more complex than it may seem. Typically it is a combination of many characteristic frequencies.

---

*For a rich look at this effect, try Robert Levine's *A Geography of Time*.[8]

No physical musical instrument can sound a pure note; there is always a *timbre* or texture to every sound. The rhythms of every situation have a texture to them.

The most basic decision-making skill is adapting to the tempo of your environment, and setting your own pace within it. It is significantly harder to become a pace-setter or pace-disruptor: somebody who can actually influence the tempo of the environment. The collection of behaviors involved in managing tempo is what people mean by the phrase *sense of timing,* and there is a lot going on beneath the sublime moments in comedy or stock trading that they have in mind when they use the phrase.

Let's start with a familiar example from a highly sensory and tactile domain: driving on a highway,[2] an experience that you can break down into four behaviors. The first behavior is merging, a task that involves managing both speed and timing. You rapidly judge the speed of traffic flow and the pattern of spacing between vehicles, and plan your acceleration so you can neatly slide into an available gap.

The second behavior is going with the flow. Cars in each lane drive along in relatively evenly spaced clusters separated by larger gaps. The speed of flow combines with the pattern of spacing to create the complex, textured tempo of the traffic. Trucks and lane-changing vehicles add to the overall rhythm. Different groups may have different local flow conditions.

Good drivers with a taste for higher speed weave through the gaps, creating a harmony, exhibiting the third behavior, pace-setting. Faster clusters may form behind them. Occasionally, a faster cluster, straddling multiple lanes, will simply pass through a slower one, creating a two-voiced fugue.

Dangerous drivers disrupt the rhythm with dissonant behaviors such as tail-gating, going too slow, or going far too fast. In traffic, legitimate practice of disruptive behaviors is limited to drivers with the right talents, reasons and social roles within the highway community: drivers of police cars, fire-trucks and ambulances. Driving decisions are strongly shaped by our intuitive mapping of people to the archetypes of highway society, such as *jerk, road-hog, maniac, old lady, cop* and *teenager,* which we associate with predictable patterns of driving behavior.

Stand by a highway sometime and listen to the raw music of traffic. Listen to the rhythms of traffic if you ever travel to a foreign country, and watch how your own anxiety levels change.

Driving graphically illustrates the four main skilled behaviors that constitute the overall skill of timing: merging, going with the flow, pace-setting and disrupting. I will introduce, at the end of the chapter, a single micro-skill, thinking in terms of temporal logic, which will help you master all these behaviors.

Emotion plays a particularly important role in timing. So let's take a look at the interplay of emotion and timing before diving into the four behaviors.

2.1 Emotion and Timing

Emotion results when you force yourself, or some part of the environment, to operate at a faster or slower tempo than it likes. To change a tempo you must add or remove energy by applying a force. The image to keep in mind is a child on a swing: the simple sort comprising a tire hanging on a rope from a tree branch. Depending on when you push, and in which direction, you can make the swing go higher, dampen it, change its direction, or change its phase. By holding the rope at different points, you can increase or decrease the frequency of the swing. The child will react to all this with delight, disappointment, relief or frustration, depending on his mood.

Dealing with any rhythmic process, such as a regularly scheduled meeting, or a daily work-life routine, involves similar dynamics. Drive up the tempo too much, and calm yields to excitement, and then to anxiety and panic (which is why driving up tempo is a big part of creating FUD, Fear-Uncertainty-Doubt, in adversarial situations; "shock and awe" is an extreme example). Add too much damping, and emotions swing the other way: calm yields to impatience, frustration, anger and finally, depressed acceptance. Our reaction to trauma or the death of a loved one – the ultimate example of a living tempo being reduced to a dead stop – goes through an exaggerated version of this reaction, the Kubler-Ross stages of denial, anger, bargaining, depression and acceptance.

The emotion of a tempo can even change with no change in pace, and no new external factors. Humans and organizations naturally crave variety and stimulation. Stay too long at one tempo, and boredom will set in.

The interplay of time and emotion is deeper than you might think. Subjective time might well be the more fundamental sort of time. Einstein's well-known quote was not a throwaway remark: "When a man sits with a pretty girl for an hour, it seems like a minute. But let him sit on a hot stove for a minute and it's longer than any hour. That's relativity." Steve Mirsky notes in *Scientific American*:[†]

> Amazingly, the pretty girl/hot stove quote is actually the abstract from a short paper written by Einstein that appeared in the now defunct Journal of Exothermic Science and Technology (JEST, Vol. 1, No. 9; 1938). Apparently, the great theoretician tried his hand, and other body parts, at experimentation to derive his simple explanation for relativity.

While it is not clear whether the relativity of time as understood in physics has anything to do with the psychological relativity of time, the latter has been studied and measured since the 1930s,[9, 10] and a great deal is now known about how we subjectively experience time, and in particular, how we systematically underestimate or overestimate durations depending on context and emotional state.

2.2 Situation Awareness

Context-switching is the process of exiting one situation and immersing yourself into another. Merging onto a highway is a small-scale example. Part of this process is what is known in the military as developing situation awareness (sometimes called "situational awareness"):

> The perception of environmental elements within a volume of time and space, the comprehension of their meaning, and the projection of their status in the near future.[‡]

Appropriately generalized, with *space* replaced by *virtual social environment,* this definition applies to most kinds of modern information work.

[†] In the September, 2002 issue, a special issue on the physics of time that is well worth digging up and reading.

[‡] This widely accepted definition is due to Mica Endsley.[11]

Situations are the background against which extended episodes of connected decision-making play out. As we'll see in the next chapter, situation awareness is a measure of the current quality and relevance of a more persistent part of your thinking called a mental model.

While developing situation awareness through a context-switch, you must usually bootstrap action while absorbing the current state of the situation. In fact, in many domains you can only develop situation awareness by acting, not by observing. In cricket, for instance, a new batsman must develop situation awareness (a process known as "getting your eye in") by cautiously playing the first few balls he faces. If he intends to dominate the bowler, a batsman will often send a signal after getting his eye in, by playing a confident and decisive shot.

Context-switching can be purely tactile and sensory (watch dancers segueing onto a dance floor), or mostly subconscious (even in verbal domains; watch a newcomer joining a conversation). There are also cases where it is a conscious and deliberate process of discovery and sense-making (think of an auditor trying to make sense of a pile of financial documents).

Situation awareness degrades if it is not actively maintained through immersion, so context-switches are an asymmetric form of information work. Preparing your mind for the *to* domain is uphill work. Letting go of the *from* domain is downhill work.

At work, for instance, it is easier to acquire the habit of making notes and capturing action items after a meeting, when the situation awareness is still strong in your mind. It is much harder to acquire the complementary habit of preparing adequately for meetings, since it is difficult to acquire the necessary situation awareness before actually entering the venue and encountering faces and voices. This asymmetry is one reason that start-up situation awareness rituals, such as agenda reviews and approval of minutes, are more common than exit rituals. Post-processing happens more naturally than pre-processing.

This asymmetry is due to the momentum of mental models, which we will cover in the next chapter.

What happens while you are developing situation awareness? Of course, you notice (or anticipate) and classify the *what, when, who* and *where* of the situation, which allows you to quickly load the right mental model into awareness, but the main action centers around developing awareness of rhythms. The world moves faster than we can hope to keep up, and we adapt by separating change into a background of predictable rhythms, and a foreground of unusual rhythms and non-rhythmic elements.§ Without this separation into foreground and background, the sheer cognitive load of keeping up with change, the process of truth maintenance, would overwhelm us. So, rather efficiently, we hold the compact belief that "the sun rises and sets everyday" and relegate that belief to the background, rather than precisely tracking the sun at all times.

Developing situation awareness is primarily the process of getting attuned to the dynamics of the background. Anything left over and unexplained is the foreground: raw material for active engagement.

Let's look at some backgrounds.

2.2.1   In the Kitchen

Let's try a context-switch. Imagine walking into your kitchen. Stand and ponder its rhythms silently for a moment.

Over years, appliances come and go. Over months, sacks of rice, tubs of margarine and bottles of dish detergent run out. Over weeks, bottles of orange juice and loaves of bread come and go. Over days, food rots, leftovers get used up, the trash can fills up, and the fridge gets restocked with fresh vegetables.

You add your own rhythms every time you cook; the kitchen comes awake when you enter. Over hours, as you cook, dishes move slowly from cabinets to countertop to sink to dish-rack, creating something of a moving wave of clutter. At the fastest end of the spectrum, knives drum out breathless rhythms on cutting boards and soups boil and bubble their way to crescendos.

Embedded in this harmony there might be the occasional moment of pure timing artistry, such as a pancake-flip. If you are too hasty, it will stick and you'll make a mess. Wait too long and it will dry out or burn.

You, as cook, eater and yo-yo dieter, change too. Over a lifetime, you acquire and cultivate skills. Sometimes you neglect and lose skills. Over years, you acquire and lose habits and learn and forget recipes. Over months, you adapt your eating to your changing exercise and work routines, experiment with different cuisines, and switch cereals. As you age and your health changes, you build in systems to deal with issues such

---

§In control theory, the background rhythms are known as natural dynamics.

as high cholesterol or diabetes. Over weeks you go grocery shopping and take out the garbage. Day after day, you ask yourself, *what shall I eat today?*

Consider emotion. Does your sink spend most of its time cluttered with dirty dishes or clean and empty? Does trash promptly go out on time or regularly overflow and stink up the house? Would you say your kitchen, right now, is calm, distracted or depressed? What about energy? Does your kitchen bustle when you cook? Does it look alert and ready for action when you are away, or lethargic? Ever been to a tired restaurant where the waiters listlessly dump insipid food on your table?

You develop a subconscious awareness of this ongoing drama of emotion, energy and rhythms, in a few seconds, every time you enter your kitchen. That's the context-switch.

2.2.2   In the Workplace

Let's look at a bigger context-switch, your experience of starting a job in a traditional organization. Reflect on your memories of that episode.

The workplace has its own base rhythms. In the United States, everything happens to the drumbeat of quarterly reporting and annual planning. Higher levels of the reporting hierarchy beat out deep, slow rhythms via large communication meetings, that act to synchronize the activities throughout the organization. A frontline team might have a heartbeat as short as a day. Software development teams that use a collaboration model known as Scrum usually adopt the ritual of the daily "stand-up meeting." Operational divisions sometimes adopt sunrise and sunset meeting rituals.

But these are just the overt rhythms. Are you sensitive to some of the more subtle ones? Do your senior managers tend to process their emails on Sunday night? Who are the larks? Who are the owls? To what extent has your workplace environment acquired a fragmented arrhythmia, created by peripatetic, mobile work-life blenders switching between work, home offices and "third places" such as cafés? How many of the emails you receive say "sent via Blackberry?" If you are one of these work-life blenders, how do you view your more stable traditional colleagues?

Non-traditional, free-agent information workers are not immune to environmental tempo.[12] The tempo of free-agent life is driven by broader patterns of activity in big cities, activity on mailing lists, and other such free-form substitutes for institutional rhythms. Becoming a free agent means becoming attuned to the tempo of a particular corner of the economy.

Release schedules of major projects drive the lives of open-source developers. Seasonal rhythms of trade shows, conferences and product releases drive the lives of independent consultants and analysts within industrial ecosystems. If you are a free-agent, how do the rhythms of your work life relate to those of the ecosystem you live in?

In your world, do emails threads typically straggle and quietly exhaust themselves in a slowing cascade of lethargic non-responses? Or do they bounce around and cascade with increasing acceleration and urgency, looking for movement and resolution with the energy of a nuclear chain-reaction?

Would you call your workplace or ecosystem a restless go-getter context, or a relaxed and laid-back one? Is there a sense of urgency? Can you read and place the culture on a spectrum between complacency and panic?

The astounding thing is that you can probably provide accurate answers to all these questions if you've been working in your current environment for more than a few weeks. That's how quickly you switch context into, and develop situation awareness of, a new workplace.

2.3 Going with the Flow

Once you are settled into a routine in a domain, and attuned to the local rhythms, your next challenge is to operate efficiently within it. For most people, this translates to a path of least resistance, which I'll call *going with the flow*. This is what most drivers do in traffic. Decision-making is hard. It pays to automate as much as you can.

Going with the flow is the smart thing to do if you've found a stable and rewarding niche, where your auto-pilot skills are in high-demand. Work comes at you self-defined (what), it is well-motivated (why), and you have the practiced skills to do it without much meta-thinking (how). The complex elements of modern decision-making – when, where and who – are reduced again to pre-modern levels of triviality. While driving in smooth traffic, these decisions are reduced to *here*, *now* and *in relation to the cars around me*.

This isn't a bad thing. Well-designed decision-making contexts are full of reasonably rational embedded systems and processes, which chug along, making vast numbers of default decisions for you, or cueing you at appropriate times, for decisions that cannot be automated. What makes us suspicious of going-with-the-flow behaviors is that they can become *ritualized* if underlying assumptions are not periodically re-examined, and the design reconsidered in light of new information. The common English idioms, *in a groove* and *in a rut* illustrate how efficiency and the potential for irrationality go together.

The apparent simplicity of going with the flow is deceptive. Through a process of externalizing mental models, we move complexity out of our minds and into the environment. We will talk about this process, as well as ritualization, in Chapter 6.

Let's revisit the kitchen and the workplace and see more of this.

2.3.1 In the Kitchen

For an example of in-flow decision-making in the kitchen, consider grocery shopping. Though cookbook authors constantly bombard us with books containing anywhere from dozens to hundreds of recipes, most of us make the same handful of recipes over and over again. Our shopping lists look roughly the same week after week. Mine usually includes tomatoes, cucumbers, bananas, bread and soy milk.

Routine begets efficiency. In *Rubbish: The Archeology of Garbage* by William Ratjhe and Cullen Murphy,[13] we learn that families that cook the same few recipes over and over generate much less food waste than those that constantly experiment with new recipes and cuisines.

Eating what is in your fridge, or cooking with whatever ingredients happen to be on hand, and within the bounds of time and energy reserves you have available, are simple examples of going-with-the-flow behavior.

2.3.2 In the Workplace

Suppose you have some new information about a competitor that you need to share. How do you proceed? Going with the flow means using established venues – your next team communication meeting for instance – to share the news. Alternately, you could set up an urgent cascade of emails that seeks out the impacted parties within your organization and triggers the setting up of an exceptional meeting.

Here's the curious thing about this choice: whether or not you decide to go with the flow will actually drive the urgency of your organization's reaction to the news more than the content of the news itself. Using a go-with-the-flow model can create a dangerously complacent reaction to critical news. Invoking exception-handling models for trivial reasons will cause over-reactions.

Of course, people and systems aren't stupid. When the content is critical enough, handling it in routine ways will set you up for sharp criticism: "did you not recognize the importance of this?" On the other hand, cry-wolf patterns of career self-sabotage happen when you cause expensive exception-handling decision-processes to kick in for stupid reasons.

2.4   Pace-setting

Pace-setting is the art of harmoniously driving the natural tempo of your environment away from its current state and towards your preferred state – slower or faster – in non-disruptive ways.

The idea that systems have natural tempos is a powerful and widely studied one. Charles Fine, in *Clockspeed*,[14] characterized different industries in terms of their differing natural rhythms. Besides the routine workplace clocks that we encountered earlier in this chapter, each industry has its own set of unique clocks, such as the product lifecycle, that help define its natural tempo. Mixing Fine's metaphor with one from computer-hardware-hacking, we can think of overclocking (making a computer chip run faster than it was designed to) as one sort of pace-setting.

To understand pace-setting more broadly, think in terms of two concepts from airplane flight: operating point and operating envelope. Individual humans and organizations are capable of exhibiting more variation in tempo than we realize. This range of variation is the operating envelope. The current tempo is the operating point.

The operating envelope is a set of well-defined safe-pace behaviors. Humans can saunter, walk, jog, run or sprint; each is a gait, a safe pace, with a particular energy consumption rate. The current dynamic state is merely one of many possible safe ones within an operating envelope.

The heart of this envelope is what we call a comfort zone, the portion of the envelope which you can traverse without significant changes in the emotional quality of the tempo.

Beyond this is the pace-setting zone. Push beyond the comfort zone and you get charged emotional responses. Running is exciting. Walking is calming. Walking very slowly and meditatively is frustrating, unless you are a monk. Push the boundary of the envelope itself, and you get dissonance and disruption, which we'll get to next.

Energy dynamics and reserves dictate how long you can sustain a pace. Most of us can walk for several hours, but sustain an all-out sprint for no more than ten seconds.

Pace-setting is about more than merely driving natural frequencies up or down. You can achieve dramatic changes in tempo without changing frequencies at all. Benjamin Franklin's classic advice from *Poor Richard's Almanack*, "Early to bed, early to rise, makes a man healthy wealthy and wise" is an example. If you've ever switched from being a lark to being an owl, or vice-versa, you have personally experienced the very different energy patterns and emotional feel of the two routines (engineers will recognize this as an example of manipulating the *phase* of a rhythm). You can maintain the same rhythm, or even slow down, but increase the energy of the rhythm, and get more done. And finally, you can manipulate the proportion of up-time and down-time in a rhythm,¶ and choose between slow-and-steady or spiky patterns of effort.

2.4.1   In the Kitchen

I discovered my favorite example of pace-setting in the kitchen by accident. One evening, I returned from an energizing day at work, looking forward to cooking, and found myself faced with a tired, dirty and grimy kitchen, with an overflowing sink. I would have to clean and revitalize the kitchen before I could get started.

I almost succumbed to the tired, buzz-killing tempo of the kitchen. But just as I was about to pick up the phone to order take-out Chinese food, something made me fight off the lethargy that was starting to descend. I attacked the sink with unnecessary ferocity, rapidly finishing up the dishes and cleaning the countertop in ten minutes.

¶In engineering, this is known as the duty cycle.

At the time, I was only able to extract a minor lesson from the experience: not to let the unpleasantness of tasks mislead you into overestimating their magnitude.

Later, I realized there was a deeper lesson: the apparently unnecessary ferocity was necessary after all. The act of bustling about – pacesetting – was necessary to get the tempo of the kitchen where it needed to be. Had I not bustled, I might well have gotten the kitchen clean, only to have psychologically exhausted myself too much to cook. I needed to get the kitchen clean *and* preserve my own buzz.

Today I have a stable pace-setting routine that I use whenever I start cooking. I move clean dishes from the dishwasher to the cabinet and move dirty ones from the sink to the dishwasher. Remember the metaphor of the child on a swing? This is an example of adding a jolt of energy to what you might call the "kitchen cycle" to lift its emotional state. Dishes in the cabinet are like an energized swing at its highest point. Dishes in the sink are like a swing that has slowed to a stop.

2.4.2   In the Workplace

For our workplace example, consider the routine matter of email responses. Here are two variants of an exchange that achieve the same outcome, but with different tempos:

Script 1

A to B (10:15 AM, Monday): Could you send me that report?

B to A (10:17 AM, Monday): Will send by COB... in a conference call right now.

A to B (10:19 AM, Monday): Ok, Thx.

A to B (9:02 AM, Tuesday): Hi, just a reminder, could you send that report?

B to A (9:03 AM, Tuesday): Oops, sorry, it slipped through the cracks. I'll send it asap. I just need to dig it out.

A to B (9:10, Tuesday): Okay thanks. I do need it today though.

B to A (9:13 AM, Tuesday): Here you go

Script 2

A to B (10:15 AM, Monday): Could you send me that report? Thx

B to A (10:19 AM, Monday): Here you go.

Script 1 took almost 24 hours to run its course. During this period, seven emails were exchanged, of which only two had any relevant information. The actual processing time for the task (finding and returning the report as an attachment) took three minutes in script 1 and four minutes in script 2.

In pace-setting decisions, the default go-with-the-flow option is often an unconscious one, while the change-the-pace option must be deliberately adopted. Consider our two scripts.

If script 1 is typical in the organization, then using script 2 would represent a speed-up pace-setting decision on the part of B. By choosing the slightly more complex, meeting-interrupting response (that took an extra minute of processing time, due to the context-switching involved), B may have driven up the typical pace of the organization slightly. If B routinely operates this way, she will act locally as an accelerator in the workplace, merely by resetting local expectations around email interactions, which could very well cascade further.

On the other hand, if script 2 is the default (as it is in many go-getter startup organizations), then B choosing to go with script 1 might have been a conscious slow-down decision to set a more leisurely, measured pace (emanating either from passive-aggressive or calming/moderating intentions).

2.5 Dissonance: Disrupting the Flow

Finally, let's consider the most complex skill: disruption. To create dissonance in artful ways, you must develop a musician's aesthetic sensibilities. This is what will turn a potentially dangerous and stupid sort of behavior into a productive one. Even when productive though, dissonance and disruption involve real creative-destruction, and have the potential to create irreversible structural changes.

In managing dissonance, keep an eye on that most subtle of all emotions: boredom. Recall what we said earlier: there is natural restlessness and variety-seeking in human and organizational personalities. In managing dissonance through decision-making, boredom is your ally.

As in music and comedy, the most sublime moments of artistry are often achieved when you judge the onset of boredom just right, and insert a smart disruption of tempo that takes advantage of the latent energy of boredom to create momentum.

We briefly encountered one example of disruption (exceptional meeting venue selection) in our earlier example of going with the flow. Let's consider a couple more.

2.5.1   In the Kitchen

Dietary changes provide great illustrations of disruption in the kitchen. A recent personal example was my switching to oatmeal as a regular breakfast food. Though my doctor had encouraged me to try oatmeal for my cholesterol, I had been resisting for months. Besides the inertia of my existing breakfast routine, I was mainly resisting because the stuff looked like unappetizing brown sludge to me.

But one morning, as I was contemplating my usual options with disinterest, my wife (who regularly eats oatmeal for breakfast) offered to make enough for both of us. I accepted out of sheer boredom with my regular regimen.

It turned out I liked oatmeal after all, and began eating it regularly. The change caused significant disruption in my routine. Among other unanticipated consequences, I found myself eating lighter lunches, and occasionally deferring coffee to mid-morning. This sort of extended impact is the reason it is so hard to adopt plausible-sounding 15-minutes-a-day panaceas for modern ills.

2.5.2   In the Workplace

Let's finish this chapter by examining a micro-level (but complex) example from the workplace: disrupting conversations by interrupting or talking forcefully over other people.

The idea that all conversations must aspire to an ideal of respectful active listening is about as interesting to the student of decision-making as a musical piece with no dissonance is to a composer. The full spectrum of healthy conversational tempos ranges from leisurely, uninterrupted taking-of-turns with active listening, through impatient and interruptive finishing of each other's sentences and violent agreement, to violent disagreement.

The example is from a documentary I once watched, about a group of Arabs and Israelis that had come together to attempt a citizen-forum stab at achieving peace in the Middle East. The most memorable part was the clash of interruption norms.

Israelis cheerfully and forcefully interrupt each other, and are comfortable with high levels of overt dissonance. The video showed the Israelis frequently interrupting, with sentences that began, "No, No. . . ." The Arabs on the other hand, were used to alternately making long, elaborate and uninterrupted speeches (a conversational norm known as *musayara*, "going together," perhaps better-suited to leisurely encounters among Bedouin tribes in the desert). Each time one of them was interrupted, he would shut up in surprise. The Israelis would then end up confused as well, since they were expecting to be interrupted in turn. The result was utter disruption of the meeting.

This behavior – interrupting and talking over others – can be so disruptive in fact, and so often the result of a disruptive temperament rather than a situational need for disruption, that it is nearly always viewed as an unpleasant and abrasive personality trait. But used judiciously, interruption and talking over others is how you, as a socially situated decision-maker, can arrest the momentum of developing groupthink and assumed consensus.

If you do not develop the thick skin to occasionally interrupt, and allow yourself to be interrupted, you will help enable pathological decision-making cultures wherever you go. Go too far though, and your thick skin will enable abrasiveness in others, or numb your ability to feel emotion (remember the ability to feel emotion is necessary for maintaining situation awareness).

2.6   Putting it Together

Consider what happens during a moment of inspired comic genius, when a great comedian breaks from his script to improvise an unforgettable gag:

1. The comedian has developed high situation awareness of the venue.

2. He is keeping the momentum developing with an easy, rhythmic flow of energy, the punch lines representing "push the swing" moments.

3. He is dominating the tempo of the evening through active pacesetting.

4. His mind manufactures a clever new line, and almost without making it to his conscious awareness, it segues into what he is saying, suddenly and productively disrupting the canned logic of the situation that has been developing.

That is what putting the different elements together looks like. The four behaviors come together within a single short window of opportunity; and risk, luck and boldness conspire to create art from craft.

This sort of magic, understandably, doesn't happen too often. When it does, the episode becomes memorable. I don't have much comedic talent, but in high school I did aspire to the status of class wit. And just once during those years, I managed a wisecrack that achieved this pinnacle of artistry. It is the only one of my jokes from those years that I still remember (it would be too tough to share the joke itself; you had to be there).

We may not all be comic geniuses, but within the scope of your particular artistic talents, you will find more opportunities to create such magic than you might think. Every time you spot such an opportunity, and take an artistic leap of faith, your boldness will increase. As it does, you will get better at spotting the opportunities.

2.7 Skill: Calendar Art

The skills in this chapter aren't directly teachable. All varieties of timing are like comic timing in that sense. They take immersion in a domain, osmosis, apprenticeship and practice to acquire. Your appetite for increasing risk will determine the extent to which you can rise above craft and get to art.

But though a refined sense of timing cannot be taught, there is a smart language you can learn to improve the sophistication of how you think about time.

Much of the conceptual apparatus around decision-making in use today cannot represent the sorts of timing ideas we've discussed. This is primarily because popular decision-making models rely on what you might call point logic: the idea that a decision is a point, a fork in a temporal road, so to speak. This limits our reasoning about decisions mostly to before-after frames.

Fortunately, a better scheme, which organizes understanding of time around intervals rather than points, was worked out by planning researcher James Allen in the early 1980s. This scheme, called interval logic, is a way of thinking about time that can raise the sophistication of how you merge, go with the flow, set the pace, disrupt it, and put it all together in your moments of inspiration. If you have managed projects, you may be familiar with Gantt charts, which are a crude and early application of interval logic.

The idea of interval logic is simple, and you should be able to work out its basics in about a minute by asking yourself this question: Given two intervals of time, of different lengths, how many qualitatively distinct relations can there be between them? It turns out that the answer is thirteen (six pairs of symmetric relationships, and one special case).

Here they are, with illustrative examples from the kitchen and workplace, and mnemonic visualizations and phrases to help you remember them.

2.7.1　Before/After

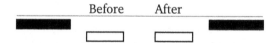

- *Kitchen Example*: Letting dough rest before baking

- *Workplace Example*: Acting after an initiative seeded earlier has incubated for a while

- *Mnemonic:* Sow and harvest

2.7.2　Meets/Met By

- *Kitchen Example*: Make the salad just before the guests arrive

- *Workplace Example*: Pep talk just before major customer event
- *Mnemonic*: Billiard ball cannon shot

2.7.3 Overlaps/Overlapped By

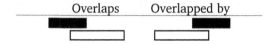

- *Kitchen Example*: Buying milk before you run out
- *Workplace Example*: Air traffic control shift handoff
- *Mnemonic*: Passing the torch

2.7.4 Starts/Started By

- *Kitchen Example*: Pre-heating oven while you make batter
- *Workplace Example*: Providing initial oversight to a new employee
- *Mnemonic*: Booster rocket

2.7.5 During/Around

- *Kitchen Example*: Cut the birthday cake after most guests arrive, and before they start leaving
- *Workplace Example*: Schedule the product demo during the coffee break at the trade show
- *Mnemonic*: Strike while the iron is hot

2.7.6 Ends/Ended By

- *Kitchen Example*: Add salt and stir just before turning off the flame

- *Workplace Example*: "Bring in the CEO to close the sale"

- *Mnemonic:* Finish with a bang

2.7.7 Equals

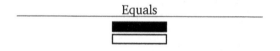

- *Kitchen Example*: "It'll take two to carry that big cake to the living room"

- *Workplace Example*: Breaking up a large meeting into break-out teams to divide-and-conquer

- *Mnemonic:* "On three: lift"

2.7.8 Examples

Once you've figured out this set of relationships, think about the role of time within the broader concept of "event," the intersection of a (possibly virtual) venue, an associated interval of time (possibly spanning many time zones), a subset of the world's social network, and an organizational context.

Can you characterize the 13 different ways to schedule a decision event in relation to a another event? Here's a worked example: a *fait accompli* can happen when you make an irreversible commitment before you hit the default venue. The *fait accompli* is an example of a pace-setting tactic.

The thirteen relationships can help you codify informal language such as "bring that up towards the end our next meeting," and become much smarter about managing your calendar.

Example 1

Once you get comfortable with the relations between pairs of intervals, you can diagram the interaction of rhythms.

If your team has a weekly meeting, should you schedule it on a Monday when situation awareness is low but energy is high, or on Wednesday when the opposite holds, or on Friday, when distraction rules, but you have the momentum of whatever was accomplished between Monday and Thursday on your side?

Should you ask your manager to drop in once a month or once a quarter to provide the right level of encouragement, boosting and feedback? Should this be an exceptional review meeting, or just before, during or after a weekly meeting? Each choice has distinct effects, and interval logic helps you see a full range of options.

Example 2

Here's a tougher application of temporal-logic thinking (you can safely skip), commonly found in industrial scheduling. In everyday life, we think of scheduling primarily in terms of start-time and end-time. But operations researchers think in terms of at least six different time variables:

- The earliest time you can start (release time),

- Actual start time

- Actual end time

- Processing time (in trivial cases, just the difference between start and end times)

- Due date/time (the time beyond which there are costs associated with being late)

- Delay (for late tasks, the gap between the due time and the finish time; the variable that determines penalties).

Of these, four are time points, one is a fundamental interval (processing time), and one is a derived interval (delay). Can you reorganize this set into two intervals, *window of opportunity* (W) and *processing time* (P) and identify the feasible set from the 13 relations?

2.7.9 Calendar Art

My goal with this skill is to get you to see your calendar – and I mean here the actual grid of dates and times you see in something like Microsoft Outlook – as a canvas for artistry. Your calendar is not an empty container. It is a landscape of invisible energy and emotion associated with all the things that are going on in your life. A good breakfast and an appreciative email leading to an energized morning could cause a great meeting. Do you normally juggle your calendar so a tough conversation is scheduled for the energized morning?

Don't get carried away by the idea though. Keep it simple; developing your artistic sense of timing, not optimal calendar management, is the point here. Focus on small dabs of artistry. Move a meeting here, try a *finished by* tactic there. If you are an owl, take a "lark day."

If you normally schedule your time in blocks of one hour, try scheduling a week using just four-hour blocks. Investor Paul Graham calls this the manager time/maker time distinction.[||] He suggests that people who build things should manage their time in four-hour blocks, while managers should use one-hour blocks.

Figure out what sort of artistic techniques you enjoy playing with, and let your dabs of exploratory artistry develop into your personal calendar aesthetic.

[||] http://www.paulgraham.com/makersschedule.html

Chapter 3

Momentum and Mental Models

There is a tide in the affairs of men
Which, taken at the flood, leads on to fortune;
Omitted, all the voyage of their life
Is bound in shallows and in miseries.
On such a full sea are we now afloat;
And we must take the current when it serves,
Or lose our ventures.

- *Julius Cæser*, Act IV, Scene 3

Situation awareness is our subjective sense of the immediate relevance and quality of an active mental model: an unwieldy, dynamic and partially coherent construct that represents our understanding of a particular class of situations. You might have one for *work* and one for *home*. Mental models are unruly. They overlap, leak into each other and form wobbly hierarchies; your mental model for *meeting* might be a small, compact one that overlaps both *work* and *home*. *Home* might contain sub-classes for situations such as *morning ritual, spend a quiet evening alone* and *host a party*.

Mental models repeatedly surface into and sink out of awareness, as you enter and exit different contexts. Learning and creative thinking create new mental models. Decisions can destroy them. They gather and shed momentum and coherence, and react among themselves. This process of Darwinian creative-destruction inside your head never stops; in quiescence and activation, in sleep and wakefulness, every mental model is always changing. It is this process that creates our inner lives.

This chapter is about momentum, a basic property of mental models that helps us conceptualize a great deal about decision-making. Momentum is the reason behaviors such as the *fait accompli*, brinkmanship, second-guessing, passive aggression and time-outs have the effects they do. Let's start with a familiar example: conversations.

Conversations are the *e. coli* of momentum-driven decision-making, and illustrate its basic properties very well. New information comes in at every step, and you use it to update your mental model of the situation and modulate the stream of what-to-say-next decisions. You steer by your overall sense of where the conversation is going, and occasionally override your conversation autopilot with a deliberate and considered response.

To get a feel for momentum, consider this classic example of a loaded conversation fragment:

> PROSECUTOR (Thundering voice): "Have you stopped beating your wife?"
>
> DEFENDANT: "Yes,...no, ...I mean ..."
>
> PROSECUTOR: "Answer YES or NO!"

Exactly what invisible load is this exchange carrying? Even sophisticated thinkers rarely go beyond the Zen answer, "Mu" (shorthand for "I reject the premises of your question"). Try *that* in court.

For an effective response – one that strengthens the case for acquittal – the defendant needs to resist the impulse to respond with fear or aggression and foreground and invalidate the hidden assumption that he beats his wife. He must then add any new assertions that might be needed, expand the space of acceptable answers, choose one, and respond.

And all this must happen very quickly because conversations move along and accumulate meaning even during interstitial silences ("pregnant pauses"). It may seem like an impossible task, but fortunately our brain manages the process rather efficiently at a partly subconscious level. You can influence this processing, and we will discuss ways to do so in the following chapters.

Consider this more situation-appropriate response to the Wife Beater question:

PROSECUTOR (Thundering voice): "Have you stopped beating your wife?"
DEFENDANT (Assertively): "Are you trying to browbeat me?"

Notice how this response refuses to acknowledge the sly framing, questions the motivations of the prosecutor, and reclaims momentum by answering the question with a question. A purely logical and emotionally neutral response such as "I don't beat my wife" would have appeared weak and defensive, while an overtly aggressive response such as "What the hell kind of question is that?" might have played right into the prosecutor's hands by exhibiting a violent temperament in front of a jury.

The four answers to the question arise from different mental models of the situation. The Zen answer, while possibly reflecting a high-quality mental model ideally suited to philosophical debates, is irrelevant to the courtroom situation. The logical answer is not emotionally charged, as a response to a wife-beating accusation should be. The aggressive answer brings in the equally irrelevant context of street-smart posturing. The answer I suggested is the sort of acceptable compromise an average adult might think up under time pressure (though admittedly, the situation is rather unrealistic).

Fortunately, that's all you need. In conversational decision-making, you don't have to get every response exactly right so long as you get things roughly right, and correct your course with subsequent responses as necessary. This is because the effects of what you say (or hear) are cumulative rather than isolated.

Keith Johnstone, in his classic work on improvisation in theater,[15] offers actors a simple principle: to improvise a scene, the only thing you really need to know is what status to play. This principle requires surprisingly few modifications when it comes to real life. At the heart of this book is the idea that only a few variables need to be managed to control a stream of decision-making. We'll encounter some of the most common variables later in this chapter.

Let's develop a few new concepts to structure our thinking.

3.1   The Vocabulary of Thought

We require a working definition of the term mental model. Here is mine:

> A mental model is a dynamic, unstable and partially coher-
> ent set of beliefs, desires and intentions held together by
> narratives that weave through the current realities, possible
> histories and possible futures of a situation.

My definition is based on the Belief-Desire-Intention (BDI) framework developed by Stanford philosopher Michael Bratman in his classic:
Intention, Plans and Practical Reason.[16] Within this framework, simple
beliefs, desires and intentions are the primitive elements that comprise
the vocabulary of thought. They are rather like the nouns, adjectives
and verbs that comprise much of the vocabulary of languages. This
choice of primitives isn't arbitrary; it turns out you can derive good ac-
counts of nearly everything else involved in decision-making, such as
planning and policies, from beliefs, desires and intentions. The relation between these three primitives and decision-making is straightforward: beliefs create or constrain possibilities, desires lead to preferences
among them, and intentions represent commitments to specific courses
of action. Each of the primitive elements can evolve in time, which is
why mental models have *momentum.*

Though Bratman's ideas can (and have) been used to build very formal and clean-edged representations of mental models applicable in artificial intelligence programs, we are interested in the far messier manifestations that operate inside our heads. Think of mental models as
being rather like the fictional Chinese encyclopedia described by Jorge
Luis Borges in his 1942 short story, "The Analytical Language of John
Wilkins":

> [In] a certain Chinese encyclopedia called the Heavenly Em-
> porium of Benevolent Knowledge... it is written that animals
> are divided into (a) those that belong to the emperor; (b)
> embalmed ones; (c) those that are trained; (d) suckling pigs;
> (e) mermaids; (f) fabulous ones; (g) stray dogs; (h) those
> that are included in this classification; (i) those that tremble
> as if they were mad; (j) innumerable ones; (k) those drawn
> with a very fine camel's-hair brush; (l) etcetera; (m) those

that have just broken the flower vase; (n) those that at a distance resemble flies.

Mental models also contain multiple narratives that loosely weave together related sets of beliefs, desires and intentions to describe the future, present and past. Some are simple, such as "I am going to medical school" or "I am this way because I grew up in a small town in a close-knit family." Others are complex what-if models of the future ("If I get a job in China, I am going to learn Kung-Fu") or speculative reconstructions of the past ("If Al Gore had won the election, the war would not have dragged on"). The general term for such narratives is *possible worlds*. Philosophers use a particular form of logic, known as modal logic, to argue about such worlds.

We will ignore philosophical nuances, and think of the relation between mental models and possible worlds as roughly the same as the one between a vocabulary and a story (which is why I use the term narrative): a vocabulary restricts the sorts of things that can be said in a story, while a story generally uses only a part of a vocabulary, though it may occasionally leak out via neologisms.

Let's look at an example: you are in the kitchen, preparing dinner. What sorts of things might we see in your mind if we took an instantaneous snapshot?*

1. Concepts such as *knife* and *apple*, things beliefs and desires are about.

2. Specific and unique real-world instances of concepts, such as "my big pan with the dent in it."

3. People-concepts, such as self, guest, role, vegetarian and individual names.

4. Assertions that are usually true, such as "apples are roundish" and "knives cut apples."

5. Vague beliefs such as "he is a finicky eater."

6. Desires, such as "I am in the mood for something crunchy tonight."

---

*We are peering into a conceptual model, of course, and what we see may or may not correlate to things that neuroscientists might find with fMRI machines that *actually* take snapshots of brains.

7. Intentions, or commitments to classes of possible worlds such as "I'll make lasagna." (It is a *class* of possible worlds because many futures might involve you making lasagna, which differ in ways that you may or may not care about, such as whether or not it rains).

8. Articles of faith, such as "healthy meals must include leafy greens."

9. True right-now beliefs such as "the big pan is in the dishwasher."

10. Premises for possible worlds like "what if one of my dinner guests is allergic to nuts?"

11. Plug-and-play beliefs such as "there is at least one vegetarian coming to dinner."

12. Awareness of dynamics such as "the water is coming to a boil."

13. Fragments of procedural (how-to) knowledge that have been activated, such as *chopping* (this can include non-conceptual elements, such as muscle memory).

14. Self-referential beliefs about the model itself, such as "I feel like I am missing something."

15. Beliefs involving numbers, such as "I have six plates" and "I have eight guests coming over."

16. Beliefs about emotions and energy, "I am frustrated with this dough."

17. Inference snippets, such as "If I don't slow down, I'll ruin this pancake."

18. Quick-reaction policies such as "If pot boils over, turn off flame immediately."

19. Beliefs about yourself, such as "I always leave everything till the last minute."

20. Informal, constantly mutating story-plans, like "I'll get the vegetables chopped and ready, get the dessert done, and step out for wine while that's in the oven."

Note that the intentions evident here aren't formal goals. They are informal, dynamic working focal points. The possible worlds in play aren't defined by Gantt charts, just informal stories.

The sheer quantity of information in a mental model provides the inertial mass that carries the momentum. Momentum comprises inertia and movement. Here is an example of how "movement" works at an elemental level.

If, at 3 PM today, I believe the big pan is in the dishwasher because I put it there last night at 10 PM, the belief is 15 hours old and somewhat unreliable, since someone else could have run and emptied the dishwasher. If I now happen to open the dishwasher and notice it is still there, the belief becomes current and refreshed – it has moved 15 hours ahead in time. If I then decide I will rinse it out and use it while making dinner between 7 and 8 PM, the big pan's future is locked in for a few hours with a fair amount of certainty, and you could say that belief has been incorporated into a possible world within the mental model that extends out 5 hours ahead of real time.

Desires evolve in similar ways: a preference can turn into an aversion; new tastes can be acquired. New desires can change the order of preference among old ones.

Finally, an intention can also evolve in time, as the beliefs and desires supporting it shift and change. An entire book could be written just about humanity's efforts to lock-in and harden intentions, once they are adopted (tactics like burning bridges and penalty clauses in contracts are among the simplest examples).

The overall movement of a mental model is not smooth. It is a staccato process comprising dozens of changes every minute to the set of beliefs, desires, intentions and possible worlds that comprise it. This is the process that we experience as decision-making. The coarse texture of this change is part of the subjective experience of tempo.

3.2   Enactments

A part of your active mental model is becoming real, or *enacted*, as time progresses. While your possible worlds might be fantasies, enactments must obey the laws of physics. Your behaviors and their consequences, which constitute the enactment, create a unique reality.[†] Your understanding of the ongoing enactment, however, will be consistent with many possible interpretations. Each is a special kind of possible world,

---

[†]Some philosophers, such as David Lewis, worry about interesting "many worlds" subtleties. These do not matter for our purposes.

made up of a *history* and an *expectation*. A history is a possible world that you claim might be *true*, while an expectation is a possible world that might *become* a history, in part as a consequence of your actions.

The relationship between an enactment and a history-expectation pair associated with it is exactly the same as the one between experimental data and a theory that attempts to explain it. Delusions, or false history/expectation pairs, are the equivalent of flawed science.

Enactments weave thought, decision-making and action in ways that you cannot hope to unravel. As one or the other of these three elements dominates, the tempo of an enactment changes. The approximate separability of certain narrow classes of enactments into regimes of "planning" and "execution" is really a matter of placing symbolic markers at points where significant and irreversible changes in tempo occur. Few enactments contain the extremely sharp tempo-change boundaries that allow you to think in terms of strongly separated planning and execution phases. The launch of a rocket is one example. But in general, the shift is more gradual, and partially reversible. The associated problem of momentum management is more subtle.

More generally, when you switch between a Plan A and a Plan B, momentum shifts from one bundle of possible worlds, to another. This is rarely a matter of turning one switch off and another one on. The clean fork-in-the-road shift is the exception; the dissipation of momentum associated with one course of action, and the cohering of another, is the rule. Enactments generally cannot be turned on a dime. Complex enactments containing a lot of momentum can be as hard to steer as an aircraft carrier.

When the momentum of an enactment is low, such as when you are sitting down with pen and paper analyzing a problem for the first time, the past and the future influence current tempo indirectly, through your conscious thoughts. When the momentum of an enactment increases, the influence is much more direct, via policies and predispositions, since you do not have time to think. To modulate this influence at higher tempos, it is important to pay more attention to the the emotional quality of the pasts and futures you are considering.

How might the emotional quality of the future and past affect the tempo of the present? Consider an example: you are driving to work and you are debating whether or not to stop at your favorite coffee shop. There are two expected possible futures: "stop at coffee shop" and "do not stop at coffee shop." Until you decide, the tempo of the enactment will be influenced by both possibilities.

If you are dreading that pending report you need to finish at work, and slightly excited by the prospect of seeing that barista on whom you have a crush, the emotional quality of the tempo you experience will be a mix of dread and excitement. The pending report might be draining momentum, while the prospect of coffee might be energizing you via anticipation. The rhythm of your driving, evident in whether or not you are riding the brakes or accelerator, and in the smoothness of your turns and lane changes, will be affected by these thoughts.

The past – your reconstructed memory of the enactment – can also affect tempo. If you ran into your neighbor on your way out of your home, and his "Hello" was a little colder than usual, part of your mind might be wondering whether he was just in a bad mood or whether he was upset with you for some reason. Both possible histories will affect the rhythms, emotions and energy patterns of the present.

Enactments begin with a theory of a situation.

3.3   Theories of Situations

Put yourself in a new situation and you will immediately surface the most relevant mental model you have available, develop situation awareness and relegate much of what is going on to the background. With whatever remains in the foreground, you will form a theory of the situation.

Let's say you've been invited to a big dinner party with hundreds of guests. Your mental model for *Party* is activated and starts gathering momentum the moment you put on your smart clothes. You enter the venue, say hello to the host, grab a vodka martini from the bar, and start idly working the room.

In one corner, you see a man talking animatedly to a woman, clearly putting on a performance, while she is listening and encouraging him with cool amusement, occasionally permitting herself a small laugh. They are facing each other. A second man stands near them, with a tense and anxious face, mostly being ignored by the other two. *Love triangle*, you instantly think. This is a theory of the situation.

To continue our analogy to language and storytelling, theories of situations provide foundations for classes of possible worlds, rather like premises do for stories. They encompass a relatively small subset of the beliefs, desires and intentions in play, woven together more tightly than the story as a whole. There is more logic to the weave, and key unknowns and conflicts are more clearly evident.

The *Harry Potter* series, for example, is based on this premise: "boy discovers he has magical powers, and leaves unhappy foster home to attend a school of magic and fulfill his destiny as the chosen one who will battle the forces of evil led by his nemesis, the Dark Lord."

Just as multiple stories can arise from the same premise, multiple enactments can arise from the same theory of the situation.

A theory of a situation is just an informal organizing pattern (either recognized from past experience, or improvised) that accelerates the development of situation awareness by providing a strong filter. New input that further refines, confirms or obviously invalidates the theory bubbles quickly to the foreground. Clearly irrelevant input is banished to the background. Elements of unclear relevance remain in limbo as a vague sense of uncertainty, creating drag, which we experience as doubt. In the case of a love triangle, the pattern in your head might be: *A* and *B* are both attracted to *C*, and *C* is mildly attracted to *A*.

Initially there's nothing more to your conscious *Love Triangle* theory of the situation than the recognized pattern, not even the names of *A*, *B* and *C*. But you feel a certain sympathy for B because he is older, shorter and uglier than *A*. You unconsciously label *A*, *B* and *C Jock*, *Underdog* and *Ice Maiden*.

Archetypes have just kicked in, and an element of your doctrine ("support the underdog") has come into play. A background desire in your *Party* mental model (such as "Have fun") has generated a fore-ground intention ("Have fun by playing the party game of trying to en-gineer a crack in an Ice Maiden's veneer of poise"). You are now in a position to influence the situation purposefully. One class of possible worlds has been labeled in your mind as representing the reality you will participate in bringing about. You have created an expectation.

3.4   Enactment in *The Party*

Returning to the party, let's say something about the apparent love tri-angle bothers you, but you can't put your finger on it. But you are the action-oriented type. A little uncertainty doesn't stop you. So you walk up and introduce yourself. You are clearly unwelcome, but you have a thick skin and you pride yourself on your ability to shove yourself into any social situation. Introducing himself, Underdog says he is an airline pilot, and you immediately spot an opportunity to make him look good: you ask for his opinion on a recent mysterious airplane crash.

Once we adopt an intention, if the outcomes of our actions mostly confirm our theory of the situation, the enactment gathers momentum. Intentions get refined and we act with increasing decisiveness. On the other hand, if our behaviors have unexpected consequences, or if the uncertainty associated with unexplained elements of the situation grows, the enactment loses momentum, until we switch to a better theory.

In this case, Underdog reacts as you expected. He launches into a thoughtful and intelligent take on the subject of the airplane crash, unconsciously turning away from you and addressing his remarks to Ice Maiden. Jock looks mildly irritated and clumsily tries to change the subject, but Ice Maiden is now interested, and steers the conversation back to airplane crashes by launching on her own story of an aunt who survived one. You note to yourself that Ice Maidens always warm up considerably once their interest is hooked. You smile, think "my work here is done," excuse yourself and move on. Your theory has proved correct, and in the process you have reinforced the utility of your archetypes. In future, you will use them more confidently.

This is a basic, no-surprises enactment; a theory of the situation was validated. The theory of the situation could also have been proved correct if you had failed: Underdog could have told the story badly and ended up looking like an idiot (leaving you thinking, "well, I tried"). Both possible outcomes are compatible with your enactment. In either case, the enactment would develop momentum and a more decisive tempo as it evolved. A tempo doodle (see Chapter 1) of the enactment would go from straggly and uncertain to decisive and strident.

But what if events had played out differently?

Let's say Underdog doesn't react in any anticipated way, but instead says with authority, "Ah, that's been beaten to death" and commands Jock to continue the story you interrupted, pointedly cutting you out. Even with your thick skin, a slight stab gets through and quickens your breath. You know you have no option but to gracefully retreat. You endure a rueful pang of "well, I read *that* wrong!"

But as you retreat you suddenly realize what has been bothering you from the start. Underdog had been looking fretfully at his watch and the door! You experience a slight sense of relief as your sense of uncertainty finds a focal point. So you instantly form an alternate theory of the situation, *Late Date*: Underdog and Jock are friends, not rivals, and Underdog is actually the dominant one of the pair. He is fretful because he is waiting for his date to show. Later, you watch a second woman hurry through the door and join the triad, at which point Underdog visibly relaxes. Your alternate theory is confirmed, but you had to suffer a slight social bruise to learn the truth.

Finally, consider a third version of the story. Let's say *Late Date* was not validated, but at some point, the three stars of our show, led by Underdog, break character completely, pull out guns and take the party venue hostage. They then proceed to systematically rob all guests and flee.

If this were to happen, your entire mental model for *Party* would be switched out for a new one, *Hostage Drama*. Within this mental model, you might explain Underdog's initial anxiety as that of a gang leader before a major robbery. If you know your Wild West history, you might switch to a new set of archetypes: *Butch Cassidy, the Sundance Kid* and *Etta Place*.

The various alternate versions illustrate the interplay among mental models, possible worlds and theories of situations, and how they relate to enactments.

The one thing all enactments have in common is the rise and fall of dramatic tension, experienced as variations in tempo, created by the changing momentum.

3.5 Decision Dynamics

As in the world of Newton's laws, steady movement in an enactment is neither interesting nor realistic, and as with Newton's laws, to change momentum, you need forces. The forces that act on mental models arise from internal stresses, other mental models, and interactions with reality-data (particularly through the addition and falsification of beliefs).

The analogy between physics and decision-making is valuable because it helps us calibrate the complexity of the underlying phenomena. Newton's laws govern relatively simple phenomena such as a swinging pendulum or the Moon orbiting the Earth. They also govern extremely complex phenomena such as the weather, avalanches and ocean currents. The laws have much greater predictive power in the simple cases, but are just as valid for the more complex cases. Real-world decision-making phenomena are closer to weather systems in their complexity.

Loose beliefs, desires and intentions create an atmosphere of drag and friction for those that have coalesced into mental models. Broad-strokes rhythms, along with the low-level movement of individual beliefs and desires, create the continuously changing background. In the foreground, possible worlds and theories of situations being created and destroyed cause the larger discontinuous jumps. The rising and sinking of mental models in and out of your awareness creates the biggest discontinuities of all.

I call the overall process decision dynamics, and you can think of it as analogous to an entire planetary climate system in your head. The climate of the Earth exhibits a rich variety of phenomena ranging from clouds and lightning to hurricanes and the jet stream. Decision dynamics exhibit a comparable range of phenomena:

1. A *fait accompli* is an irreversible commitment to a course of enactment before a stakeholder has a chance to contribute. When I pull a *fait accompli* on you, I suddenly close off other options you thought were open. Your alternate mental models or possible worlds die a sudden death, taking your psychological investment – the momentum – with them.

2. *Brinkmanship* is the calibrated letting go of control of a consequential decision (classically, the decision to pull a nuclear trigger). In a nuclear face-off, when one government decides to move troops closer to the border or raise the alert level, the other side's mental model for *Nuclear Holocaust* acquires momentum through the decisive movement. New theories of situations and possible worlds, such as "rogue field commander with a hair-trigger temper" come to the forefront.

3. *Procrastination* occurs when we delay a context switch by adding more momentum to our current mental model, thereby making it harder to displace. Cleaning and organizing your apartment to avoid working on your dissertation is easier than enduring the pain involved in surfacing an incoherent mental model charged with negative emotions into conscious awareness. Procrastination

usually takes the form of *displacement*, which allows us a safe outlet for the emotions we are trying to avoid. Cleaning and organizing your apartment helps relieve the growing anxieties associated with *not* cleaning and organizing your dissertation material. Within limits, this is often a *rational* reaction. The displacement activity can often serve as a warm-up for a tougher task. Deluding yourself that you are mentally prepared, and diving into a demanding activity without a warm-up, could lead to unpleasant outcomes that create a vicious cycle of increasing aversion.

4. *Second-guessing* works by forcing someone to reverse acts of destruction. If I delegate a decision to you, you quickly spin up a set of relevant mental models, work to get a lot of momentum into them, pay the cost of killing many possible worlds, and experience the relief of a lightened load to carry. Then, by second guessing, I suddenly demand that you resurrect dead models, so I can validate or override your decision. Next time, you won't put so much momentum in to begin with.

5. *Passive aggression* works by fragmenting and dissipating momentum. Your mental model identifies something as relevant, say a fax to a customer that ought to clinch a sale. Your administrative assistant (who hates you), sees the email with URGENT!!! in the subject line at 5:03 PM, but goes home anyway, and innocently exclaims the next day, "Oh, I didn't see the email till this morning." The reason this is so effective is that the passive aggressor at once kills one of the positive possible futures you had anticipated (*Sale!*), so you experience loss, and simultaneously forces you to start spinning up several new possible worlds to figure out their unclear intentions. The coherent momentum of a single, positive mental model has been dissipated and fragmented.

6. *Time-outs* are particularly interesting. Mental models don't include emotions (though they may include beliefs about emotions, like "I am angry"), but their momentum is coupled to emotions (you remember where you were during 9/11 because the emotions gave your throwaway mental model of your surroundings enough momentum for life). Waiting drains emotions from situations and associated mental models, thanks to drag from other mental models and emotionally neutral stimuli. Powerful emotions subside. Time-outs are not always reliable though. Waiting might actually give you time to reason through and discover more

things that upset you, leading to a "getting angrier and angrier" effect.

7. *Defaulting* is a phenomenon that is unique to systems that combine defaults (incumbent decision settings) with pre-programmed commitment patterns. If you don't keep up and over-ride where necessary, you will default to outcomes that you may or may not like. The metaphor of the boiled frog is often evoked for extreme examples of defaulting: a frog in a slowly heating pot of water can get boiled alive because it does not notice the gradual increase in temperature in time. Defaulting is a particularly important phenomenon when automated and computerized processes are involved.

8. *Building mindshare* is a collective-momentum building tactic in organizations, especially for changing system defaults. Simply holding meetings, delivering presentations with examples, what-if stories and (in the best instances) prototypes and demonstrations, creates momentum. You do not need to drive overtly towards a decision.

There are many other phenomena that illustrate the dynamics of mental models, some captured in clever phrases that implicitly refer to momentum and progressive commitment: *in the fullness of time, strike while the iron is hot* and *that ship has sailed*, are classic English expressions. A particularly literal one is the Hindi proverb *eet ka jawab pathar se dena*, (literally, *to reply to a brick with a rock*; i.e. reacting to provocation with escalation). There are many more phenomena without such evocative descriptions. We deal with momentum intuitively, not just through categories of named behaviors.

3.6 Dialectics

A useful theory of decision dynamics should help us understand and predict the effects of the three forces that act on a mental model: interaction with reality, interaction with other mental models, and internal stresses. We should expect the explanatory and predictive power of the theory to be closer to meteorology than billiard-ball physics.

Constructing a theory of these complex phenomena requires the idea of dialectics. A dialectic is a framework for understanding how social processes create and apply notions of truth through debate. Socratic, Hegelian, Vedantic, postmodern and Zen models of argumentation are examples.

If you think of multiple mental models within your mind as conversation partners, dialectics give you a way to think about individual, inside-your-head decision-making as well (another use of Minsky's "Society of Mind" metaphor that we encountered in Chapter 1). If you include Nature as a potential conversation partner, you get to ideas such as Marx's dialectical materialism, Joseph Schumpeter's idea of creative destruction and Kuhn's theory of paradigm shifts. The two notions of Jihad in Islam (one signifying internal struggles of the mind, and the other signifying worldly "holy war") are both dialectics. Dialectics give rise to theories of science, engineering, culture, religion, politics and economics.

Processing this cartload of philosophical ideas is clearly beyond the scope of this book. Fortunately, we will not need to. The main reason you should be aware of these ideas is calibration. Do not look for theories of swinging pendulums or moons orbiting planets. To understand decision dynamics, you will need to study things as complex as the weather, and as subtle as philosophy. While pendulum-level decision-science theories, such as game theory, can be useful in understanding very specific situations, using them as generic theories of decision-making can lead to a false sense of clarity and security.

3.7   Archetypes and Doctrines

You will rarely need to deal with your mental life at the level of BDI models, possible worlds and dialectics, but once you develop a basic understanding of those ideas, you can deal with more practical constructs a lot more thoughtfully. What we really use everyday are aggregate constructs (just as meteorology deals with somewhat loose constructs like *tornado* or *cumulus cloud* rather than individual air molecules). The two most important practical constructs are archetypes and doctrines, which we encountered informally earlier in this chapter.

Mental models acquire and shed transient momentum in specific situations, as you enter and exit them. But there are parts that persistently accumulate momentum through a lifetime. A lot of this momentum is tied to the one common feature of all your experiences: you. Your mental model of yourself is a self-archetype. More generally, your mental models of people are archetypes.

Archetypes can be deeply thought-provoking constructs. The Greek philosopher Archilochus, for instance, made a cryptic remark that has occupied other philosophers for centuries: "the fox knows many things, the hedgehog knows one big thing." Isaiah Berlin, in a famous essay analyzing the work of Leo Tolstoi,[17] applied the fox and hedgehog archetypes to artistic styles, and concluded that hedgehogs look for a grand, unified world-view, while foxes are happy with a highly fragmented one.

Whatever your self-archetype, it is the construct through which your sense of self inexorably gathers momentum through life.

Archetypes are your mental models of people. *Fox* and *Hedgehog* are particularly thought-provoking ones, but don't look for a taxonomy or more/less fundamental types. Though you can bring some discipline to your understanding of archetypes, they are essentially artistic rather than analytical constructs (keeping Borges' Chinese Encyclopedia in mind is helpful here as well). Some, like those of Freud and Jung, acquire modest amounts of rigor. Eric Berne's bestseller, *Games People Play*,[18] is a study and catalog of interpersonal interactions based on Freudian archetypes, while the popular Myers-Briggs test is based on Jungian archetypes.

Some archetypes, like *Soccer Mom*, exist for a while as abstractions in popular culture, while others get personified, like *Joe the Plumber* (working class American) and *Britney Spears* (pop-icon-train-wreck). Still others, such as *Lady Macbeth* or *Sherlock Holmes*, are powerful enough to influence everyday language, even though they only exist in fiction. ‡

But the most interesting archetypes are very local: the informal and implicit models we all develop of ourselves and those around us. In an argument, when you fume, "why do you always have to ...," an informal archetype is at work.

Archetypes are critical. We couldn't function without them, and conversations again illustrate this well. In the courtroom situation with which we started this chapter, the defendant might be thinking of himself as *Victim* or *can-beat-the-system guy* and might view the prosecutor

---

‡I made my own modest contribution to the world of pop-culture archetypes in a series of popular posts on my blog, based on the TV show, *The Office*, and a drawing by cartoonist Hugh MacLeod. Search online for *The Gervais Principle*.

as *Machiavelli* or *truth-seeker*. If we didn't have these efficient models in our head, conversations would be interspersed with silences of minutes, rather than tenths of seconds. Archetypes operating in conversations modulate the tempo of our decision-making: if you believe you are *can-beat-the-system guy* and the prosecutor is *Machiavelli,* your conversation tempo defaults might get set to *energy: high, emotion: assertive, rhythm: quick.* Archetype assessments and tempo-control variables are part of the small set of variables we mentioned in the beginning of this chapter, which allow us to control real-time decision-making more efficiently.

Archetypes give rise to doctrines. Doctrines are basic sets of beliefs and desires relevant to decision-making. They are particularly relevant to momentum management. Doctrines can be closely identified with associated archetypes. Your most stable beliefs, the ones that actually modulate your behavior, aren't about life purposes; they are about momentum management. You are more likely to switch religions than to switch from an impatient to a patient temperament.

The mystic Punjabi poet, Kabir (1440-1518) was particularly fond of offering pithy doctrines in the form of couplets about tempo and momentum:[§]

> What you would tomorrow, do today,
> what you would today, do now.
> In a moment, the world may end,
> How will you finish then?

and in a rather contradictory vein

> Slowly, slowly, oh mind, slowly the world evolves.
> The impatient gardener may flood his orchard,
> but the fruit will ripen only in season.

An article about auto-racer-turned wine-maker Randy Lewis, in the December 2008 issue of *Autoweek*, provides a great example of the stability of doctrines:

> The slow deliberate pace of winemaking and the fast, risk-taking spirit needed for racing are indelibly intertwined for

---

[§] Readers familiar with the poetry of Kabir will recognize these as loose translations of the famous *kal kare so aaj kar* and *dheere dheere, re mana* couplets.

Lewis...his and [his wife] Debbie's wines reflect their personalities. He's a risk taker, and he's driving pretty fast...not sitting still.

The relation between archetypes and doctrines extends to organizations as well. At the level of nations, for instance, Alfred Thayer Mahan noted[6] that through profound political changes spanning several centuries, the British, Dutch and French naval doctrines continued to reflect their respective national personalities.

Personally, when thinking about the ideas in this book, I am most attracted to Archilocus' approach: relating archetypes to animal symbols.[¶] Animals often exhibit exaggerated and hard-wired versions of characteristic human behaviors, and spark much useful introspection.

My favorite examples of animal archetypes are from the British children's classic *The Wind in the Willows*.[19] Another British writer, Philip Pullman, has written an entire science-fiction/fantasy series, *His Dark Materials*,[20] based on an alternate universe where individuals are associated with explicit animal archetypes called daemons. Mythologies around the world associate supernatural beings with characteristic animal symbols (for example, the "familiars" found in European lore about witchcraft. The major gods in the Hindu pantheon are associated with animal companions called *vahanas* or "mounts" that symbolize something about their natures.)

So let's review some examples of doctrines and associated archetypes. Two cautionary notes:

First, keep in mind that archetypes are an artistic thinking tool, not a dictionary. If you don't feel the urge to mess with this list, you probably shouldn't use it at all.

Second: every universe of archetypes emphasizes some aspects of human behavior at the expense of others. Be aware of what your favorite doctrines and archetypes leave out, and avoid locking yourself into limiting self-perceptions, or consigning others permanently to pigeonholes.

¶Other approaches are possible of course. I once met an executive coach whose clientele included a lot of Air Force personnel. His stock-in-trade was aircraft archetypes.

3.8 Common Doctrines

3.8.1 Ready-Fire-Aim (RFA)

Illustration: The passionate poetry teacher, John Keating, played by Robin Williams in the movie Dead Poets Society

- Motto: *Carpe Diem*

- Momentum axiom: get ahead and stay ahead of the world

- Characteristic emotion: passionate joy

- Characteristic energy state: aggressive and charging

- Characteristic rhythm: fast

- Characteristic belief: it is better to ask for forgiveness than permission

Archetype: Pit Bull

3.8.2 Resistance is Futile (RIF)

Illustration: The robotic assassin from the future, in the movie The Terminator, played by Arnold Schwarzenneger

- Motto: I'll be back

- Momentum axiom: be an irresistible force

- Characteristic emotion: ice-cold calm

- Characteristic energy pattern: relentless and overwhelming

- Characteristic rhythm: slow-medium

- Characteristic belief: what doesn't kill me only makes me stronger

Archetype: Elephant

3.8.3 Move Mountains (MM)

Illustration: The character of Andy DuFresne, played by Tim Robbins, in the movie The Shawshank Redemption

- Motto: Rome wasn't built in a day

- Momentum axiom: one step at a time

- Characteristic emotion: reserved, low-reacting intensity

- Characteristic energy pattern: gentle and steady

- Characteristic rhythm: slow

- Characteristic belief: I can do anything I set my mind to

Archetype: Beaver

3.8.4 Enjoy the Ride (ETR)

Illustration: The Hakuna Matata episode in the animated film, The Lion King, when Simba learns the joys of a relaxed lifestyle

- Motto: *que sera sera*

- Momentum axiom: let the world carry you along

- Characteristic emotion: cheerfulness

- Characteristic energy pattern: languid and playful

- Characteristic rhythm: slow

- Characteristic belief: don't take things too seriously

Archetype: The Water Rat, the character from The Wind in the Willows who lived an easy-going life on the riverbank

3.8.5 Do Your Duty (DYD)

Illustration: The character of Aragorn in The Lord of the Rings

- Motto: pay your dues

- Momentum axiom: pull your weight

- Characteristic emotion: troubled, bordering on neurotic

- Characteristic energy pattern: as demanded by the situation

- Characteristic rhythm: as demanded by the situation

- Characteristic belief: life ought to be fair

Archetype: The Mole, the central character in The Wind in the Willows, who we meet spring-cleaning his house on the first page

3.8.6 Next Shiny New Thing (NSNT)

Illustration: The character of Max in the movie Rushmore, a schoolboy who gets excited about one extra-curricular pursuit after another

- Motto: I have to have that!

- Momentum axiom: what are we waiting for?

- Characteristic emotion: excitement alternating with boredom

- Characteristic energy pattern: impulsive spikes

- Characteristic rhythm: excited alternating with lethargic

- Characteristic belief: variety is the spice of life

Archetype: Toad in The Wind in the Willows, who gets all the other characters into one adventure after another

3.8.7 Like a Rock (LAR)

Illustration: Gandalf in The Lord of the Rings

- Motto: stand your ground

- Momentum axiom: be an immovable object

- Characteristic emotion: determination

- Characteristic energy pattern: matches the force being opposed

- Characteristic rhythm: matches the rhythms being opposed

- Characteristic belief: the world is a dangerous place

Archetype: The Badger, the strong, reclusive character in The Wind in the Willows who saves the other characters when they get in trouble

3.8.8 Prepare to be Assimilated (PTBA)

Illustration: The character of mathematician John Nash, played by Russell Crowe in the movie A Beautiful Mind

- Motto: my way or the highway

- Momentum axiom: why stop now?

- Characteristic emotion: self-absorption bordering on paranoia

- Characteristic energy pattern: large reserves, steadily unleashed

- Characteristic rhythm: the beat of an internal drum, divorced from environment

- Characteristic belief: premature optimization is the root of all evil

Archetype: Archilochus' Hedgehog

3.8.9 Maybe This Will Work (MTWW)

Illustration: Leonardo Da Vinci, Mozart

- Motto: today is the first day of the rest of my life

- Momentum axiom: let's rock and roll!

- Characteristic emotion: optimism

- Characteristic energy pattern: quick recharge

- Characteristic belief: life ought to be fun

Archetype: Archilochus' Fox

3.8.10 Die Another Day (DAD)

Illustration: Tyler Durden, played by Brad Pitt, in the movie Fight Club; Camus' Sisyphus

- Motto: hit me with your best shot

- Momentum axiom: be a force of nature

- Characteristic emotion: irony

- Characteristic energy pattern: whirlwind and suicidal

- Characteristic rhythm: manic

- Characteristic belief: there are no absolutes

Archetype: Homo Sapiens

3.9 Skill: Archetype Impressions

If you want to practice and acquire the skills associated with the ideas in this chapter, Keith Johnstone's *Impro*[15] should be on your bookshelf. Here, I can only offer a small glimpse of the possibilities.

While broad-strokes archetypes like the ones in the previous section can be good starting points, both for introspection and for analyzing others, the real skill to be learned here is the construction of unique archetypes: mental models of single individuals. Building a unique self-archetype is a complex project where sadly, you are alone, but your mental models of others, even when unique, can be considerably simpler.

But to build even simple unique models of others, you need data about their characteristic patterns of behavior.‖ One of the best ways to find data is to notice and occasionally mimic repeated phrases and pet assertions used by those who are important in your life. You probably already practice this behavior – with exactly the wrong patterns, and in exactly the wrong situations. We tend to mimic others in derisive ways, to undermine their ongoing behavior during situations like fights, and we generally pick behaviors that arise from weaknesses. While this may help you win on the rare occasions that you must actually fight another person in zero-sum ways, generally this is counterproductive.

Instead, try to notice the positive patterns of behavior you see arising from people's strengths. This isn't just me preaching a positive-values philosophy. If you are being smart and surrounding yourself with effective people, this will actually lead to more accurate archetypes, since their strengths are likely more developed than their weaknesses.

It is hard to conscientiously observe and analyze people like this. Most of us are just not that interested in others. So here is the process you must practice:

1. Notice repeated phrases and assertions, like "Are we ready yet? Let's GO!"

2. Anytime you notice an instance, reinforce your observation by scribbling a note, or repeating it silently.

3. Occasionally, either in jest, or in relevant situations where your subject might trot out the phrases, mimic-repeat the phrase. You've achieved timing artistry if you ever manage to say it the exact same instant they do. You've achieved a "finish each other's sentences" level of mind-meld with your target.

4. You can then move on to exhortation: using their pet phrase when they themselves hesitate, as in "What do you say, Let's GO! Right?"

‖ Jennifer von Bergen, in *Archetypes for Writers*[21] defines an archetype as an "imprint of a pattern of human behavior."

5. When you're really comfortable with the pattern, you can use it in a "hang by own petard" tactic: persuading people using their own favorite assertions. Obviously, this can be a form of manipulation, but in the best of cases, you can tactfully help others be consistent and lock-in their positive behaviors firmly.

Mimicking someone right in front of them is a must if you want to actually validate that you are observing a valid pattern. If it is a positive pattern, the other will appreciate it being noticed, or if they are unaware of it, be grateful to you for helping them with some self-awareness. If you made a mistake, your subject will challenge your perceptions.

Once you learn enough of another person's patterns through mimicry, you may be able to find a match to one of your favorite broad-strokes archetypes. Perhaps your boss is approximately a fox or a hedgehog. If you find that your archetypes are getting very richly defined, and fit no template you know of, perhaps you should write a novel and create memorable archetypes for others to learn from. I still wonder where Dickens came up with Mr. Micawber.

We have our characters and basic tools of story-telling. In the next chapter, we will learn to tell stories.

Chapter 4

Narrative Rationality

> "What is the greatest wonder in the world?" asked the Yaksha. "Every day, men see creatures depart to Yama's abode and yet, those who remain, seek to live for ever. This verily is the greatest wonder," replied Yudhishthira.[22]
>
> – *The Mahabharata*, trans. C. Rajagopalachari

Narrative rationality is an approach to decision-making that starts with an observation that is at once trivial and profound: all our choices are among life stories that end with our individual deaths. Surprisingly, this philosophical observation leads to very practical conceptualizations of key abstractions in decision-making, such as *strategy* and *tactic,* and unique perspectives on classic decision-science subjects such as risk and learning.

You are going to need these new concepts and perspectives, because in this chapter, we will grapple with the sorts of ambiguities and complexities that stress, and ultimately break, the cleaner concepts that work in simple situations, such as projects, plans and goals.

We are about to enter truly dangerous territory. Narratives, especially cradle-to-grave life narratives, are powerful, unavoidable, and dangerous tools. The dangers led one of my favorite writers, Nicholas Nassim Taleb to argue, in *The Black Swan*,[23] that all narrative thinking should in fact be considered flawed. An entire chapter of the book is titled *The Narrative Fallacy*. Other thinkers in the decision-making tradition that Taleb represents (behavioral economics) adopt an even stronger position against narrative thought.

The critiques are valid, and are based on the observation that thinking in terms of stories leads to all sorts of biases. What critics miss, however, is that there is no such thing as non-narrative thought, free of possible worlds and ongoing enactments. There are *always* multiple narratives at work, framing our perceptions, memories, active thoughts, decisions and actions.

The idea that there is always a narrative at work is one aspect of the decision-making philosophy in this book, which is a *situated* decision-making philosophy. It is based on the assumption that there is no meaningful way to talk about specific decisions outside of a narrative frame and a concrete context, any more than it is possible to talk about physics without reference to a specific, *physical* coordinate frame (the basic idea in Einstein's relativity). For this reason, I think of situated decision analysis as a sort of theory of relativity for rationality.

By relative rationality, I mean that there is no privileged, narrative-independent model of decision-making that can be labeled absolutely rational. Models of rationality lie *inside* the mental models and narrative contexts that operate by them. This sort of philosophical relativism is a strong position, and leads to interesting consequences.*

For us, it is sufficient to note that narrative rationality isn't a single idea; it is the idea that there are many useful, local and temporary rationalities. The immediate benefit of this approach is that it allows us to take stories seriously, and make use of notions of truth other than scientific empiricism. We can meaningfully ask and obtain value from questions such as "Has he lived a fulfilling life?" or "Is this movie funny?" that only make sense within particular narrative contexts.

The dangers of narrative remain, of course. But the best we can do to defend against them is to add an element of ironic skepticism and systematic doubt to our narrative imagination. This is tough precisely because we can only see the world through a mental model in the first place.

4.1   Deep Stories and Liminal Passages

Though we started with thoughts about death and truth, this is not a chapter on metaphysics. We are primarily concerned with the *timing*

---

*This position is roughly what is known in philosophy as methodological solipsism, often employed in skeptical arguments.

of thoughts around mortality and the meaning of life, not their content. Consider this classic "meaning of life" soliloquy from Shakespeare's *Macbeth*:

> To-morrow, and to-morrow, and to-morrow,
> Creeps in this petty pace from day to day,
> To the last syllable of recorded time;
> And all our yesterdays have lighted fools
> The way to dusty death.

This nihilistic passage occurs towards the end of the play (Act V), right after the death of Lady Macbeth. Like Macbeth, you and I turn to metaphysical questioning in the brief interludes, known as liminal passages, between the waning of one important life story and the waxing of another. You may remember an evening of introspection between high school and college, or between college and your first job. Between liminal passages, we live through a special kind of enactment, which I will call a *deep story*. Unlike ordinary enactments a deep story is an episode of creative destruction that is significant enough to transform you. The transformation is a rebirth of greater or lesser magnitude.

Deep stories represent the largest scale at which we enact meaning. Liminal passages chop up our lives into deep stories, creating the fundamental tempo of life. The character of this fundamental tempo is what people have in mind when they make assertions like "I've always had to struggle in life," "I feel like I've finally arrived" or "I feel like I have always been a spectator on the sidelines."

4.1.1 Discovering Deep Stories

How do you find deep stories in your own life or in the lives of others? One way is to look between instances of the telltale symptom of extended existential musing. Another way is to look for acts of significant creative destruction. But perhaps the easiest way is to ask questions about motivation, until you get vacuous answers.

This works because the explanatory basis of last resort (or *frame*), for most intentional human behaviors, is a deep story. Here is an example set within *College,* a commonplace deep story between the liminal passages of leaving high school and starting your first job. Imagine a college student rushing to class one morning at 8:30 AM. If you were to repeatedly ask the student, "Why?" the explanations would become increasingly significant up to a point:

> Q: Why did you rush out at 8:30?
> A: To make my 9:00 AM Mandarin class.
> Q: I thought you liked to blow off early classes and sleep in?
> A: I like this one, I might major in linguistics.
> Q: Why linguistics?
> A: I like to travel. I might want a job involving travel after I graduate, especially to Asia.
> Q: Why not just quit college and bum around the world?
> A: Well, you need a degree for the well-paid international careers, and the good opportunities are now in China.

At this point, you've hit the boundary of the deep story. Were you to persist beyond this, the answers would likely be increasingly banal (without revealing specifics like *China* and *Mandarin*) and doctrinal (*Well, money is important*). Beyond the next liminal passage, even that over-motivated student who seems to have her life planned out will likely not be able to provide informative answers. The little details that reveal the presence of a deep story will be missing.

Incidentally, it is no accident that we hit banality with just four questions. Sakichi Toyada, the founder of Toyota, and an industrial engineering pioneer, developed the "Five Whys" heuristic, which is used today in formal root-cause-analysis investigations. The heuristic simply suggests that it usually takes about five *why* questions before you get to the root cause of a problem. Of course you can continue asking beyond that, but as we saw, the answers are likely to be generalized and trivial beyond a point.

This is only partly due to the uncertainties involved in planning too far ahead. The bigger reason is that we all subconsciously realize that our entire meaning-of-life equation could change with the next expected (like graduation) or unexpected (like paralysis following a car accident) liminal passage. We know from experience that these passages have a unsettling way of transforming us and wiping our meaning-of-life slate clean. We expect to be transformed by significant life experiences and know that we cannot predict the behavior of the person who emerges at the other end of a deep story. Adopting intentions that apply to a future, transformed you, makes no more sense than micromanaging the future adult life of your child.

Deep stories, liminal passages and the fundamental tempo together provide the foundations for narrative rationality, which is the capacity to experience time and see the world through stories. Let's see what the ability can do for us that simpler decision-making models cannot.

4.2 Enactment and Storytelling

The first elements of what I call narrative rationality were proposed by Carl von Clausewitz in his treatise *On War*.[5] These concepts were validated repeatedly by famous decision-makers from Napolean to Eisenhower, but they are not the dominant ones today. In his excellent modern treatment of Clausewitz' ideas, *Strategic Intuition*,[24] William Duggan notes that our modern ideas of decision-making derive from the ideas of Clausewitz' contemporary, Antoine-Henri Jomini.

Jomini's process-oriented approach (represented today by thinkers in the business realm such as Michael Porter) is what I call *calculative rationality*. Its dominance is clearly illustrated by this quote from *The Behavioral Revolution*, an article by David Brooks in the *New York Times* (Oct 28, 2008):

> Roughly speaking, there are four steps to every decision. First, you perceive a situation. Then you think of possible courses of action. Then you calculate which course is in your best interest. Then you take the action.

Brooks is describing textbook calculative-rational behavior. The first step is what we have called developing situation awareness. The second and third steps together constitute option generation and planning. The fourth step is what we usually call execution. Note a critical missing piece: there is no mention of mental models. Since perceiving a situation requires a mental model, calculative rationality generally assumes the existence of an appropriate mental model, rather than encompassing the creation of one. If you adopt Porter's approach to business strategy, for instance, you might use his "five forces" model of competition as your mental model.

This simple view is not wrong, it is just limited to simple situations that fit one or more of your existing mental models very well. In complex situations, planning based on such models is merely a training exercise to sample the space of possible worlds, get a sense of the complexities involved, and calibrate your responses appropriately. This is what Eisenhower meant when he said, "plans are nothing, planning is everything." Marc Andreessen, creator of the Web browser Netscape, described the idea more clearly:

> The process of planning is very valuable, for forcing you to think hard about what you are doing, but the actual plan that results from it is probably useless.

While few modern decision-making paradigms adopt the simplistic view of plans as prescriptions for execution, most assume it as a foundation, and add various extensions (an example is the inclusion of real-time feedback as a means to partially compensate for deficiencies in pre-execution information).

Narrative rationality is based on a very different foundation, the structure of stories. My definition is as follows:

> *Narrative rationality is the ability to think, make decisions, and act in ways that make sense with respect to the most compelling and elegant story that you can improvise about a developing enactment.*

The alert among you will notice that I make no reference to reality in this definition. This leaves the door open to potentially *irrational* behavior. If you only demand that your actions make sense with respect to the story you are telling yourself, you could be acting rationally within a completely delusional story.

The reason is simple: extended and complex enactments cannot be fully tested against reality *a priori*, or even *a posteriori*, as controversies in the study of history demonstrate. You must expend significant time and effort before a meaningful test of the validity of the narrative becomes possible, and often, the interpretation of the results will still remain ambiguous. Narrative rationality allows you to structure your behaviors meaningfully even when feedback is impoverished, delayed and ambiguous.

Familiar, everyday stories derive in various ways from enactments, deep stories in particular. They range from straightforward narration of a default "history" to imaginative fictional reconstruction and symbolic representation. Here are some common kinds of ordinary stories, all of which echo the structure of deep stories.

1. The narrative of a slide presentation or a speech

2. The narrative of a project in a grant proposal

3. The narrative associated with a product or corporate brand

4. The expository narrative used by a teacher

5. The history of a country as set out in schools textbooks

6. A *user story* where the narrative describes human interaction with a technology system (especially a software system)

7. A *concept of operations* (CONOPS), a slightly dated term that refers to narratives illustrating patterns of military operations

8. The purely formal narrative structure of a piece of instrumental music (which might be wholly or partially improvised in real time, as in jazz or raga music)

9. The ritual structure of a wedding (which, in the scripted Western variety, can require a rehearsal)

10. The narrative in a work of non-fiction, such as an academic paper

11. An ethnography detailing the experiences of an anthropologist while immersed in a culture (often called a "thick description")

12. The grand narrative of a culture, such as Manifest Destiny for the United States in the 19th century

13. The narrative in a complex video game

14. The narrative in a telemarketer's phone pitch

15. And of course, the most familiar stories: those that drive fiction, all the way from three-panel comic strips to epic movie trilogies, and mold-breaking literary efforts

What is common to these familiar sorts of stories is an overt structure that mimics the experienced structure of a deep story. The simplest description of this structure is known as the Freytag triangle.[25]

4.3   The Freytag Triangle and Its Derivatives

Several models have been introduced in this book so far, ranging from models of time (interval logic) to models of people (archetypes) and decision-making stances (doctrines). I've even introduced a meta-model of how we build and use models (mental models). This chapter explores the most controversial models yet: models of stories.

Recall that mental models are constructs that represent our understanding of classes of situations that are more similar than not. Deep stories, by contrast, are enactments that create *sui generis* mental models applying only to one significant *new* situation. Simple enactments can unfold on the basis of an existing mental model. Deep stories, however, would be impossible without narrative rationality. They require us to continuously improvise the background story while acting. This means constructing the mental model as the enactment unfolds (learning in the most general sense).

During the construction phase, situation awareness is very poor for an extended period, and is experienced as the disorientation characteristic of early phases of learning. By contrast, the context-switching period of a normal enactment is short, and managed subconsciously. No active learning behaviors are required.

The significance and uniqueness of an enactment depend on your role in it. A priest may conduct hundreds of weddings, and to him, *Wedding* might be a routine enactment where his role is always roughly the same. From the point of view of the bride or groom, the wedding may be a deep story, bookended by the liminal passages of first meeting and post-honeymoon reality shock, with a climactic peak occurring at the actual ceremony.

Even if we subsequently live through similar experiences, and apply what we have learned, we never relive the first experience. Liminal passages, correspondingly, are *sui generis* context switches between the unique mental models created by deep stories. They are marked by their singular content: existential musing. The only constraint we will impose on our structural model of a deep story is that it begin and end with a liminal passage.

The simplest pattern that satisfies this constraint is the rise-fall structure of simple stories, such as fairy tales, which begin with one stable state ("Once upon a time..."), proceed through a climax, and recede to a different stable state ("...happily ever after"), where something has changed (the prince and princess are united, for instance). This basic structure is captured in a classic model known as the Freytag triangle.

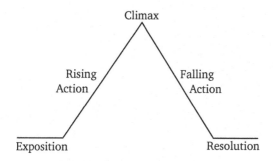

The Freytag Triangle

Storytelling is a uniquely human mode of cognition, but shorn of its meaning, this characteristic rise-fall structure seems rather like the biological stress response that is common to most animals. It is reasonable to speculate that the pattern *is* in fact the stress response, modulating cognition (though my approach does not rest on this conjectured biological justification). Later, we will informally interpret the y axis as entropy, or a measure of the disorder in your developing mental model.

Many enactments have this rise-fall structure. When stories are insignificant, like *shopping trip*, this rise-fall structure is so muted that we do not experience much of an emotional response. The bookend stability and relief experiences provide some psychological release, but they are not liminal passages. There is a peak or climactic moment when making your payment and checking out, but it does not involve much drama or intense emotion. At that point, the story merely changes from uphill to downhill.

The basic Freytag triangle, however, is too simple to exhibit all the characteristic features of a deep story. If you add certain ideas derived from a model called the monomyth, developed by Joseph Campbell in his controversial classic, *The Thousand Faces of the Hero*,[26] you get a slightly more complex diagram with two peaks and a valley in between. I call it the Double Freytag triangle.

The Double Freytag triangle represents a tractable level of complex-

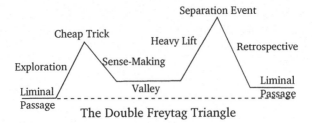

The Double Freytag Triangle

ity. It is neither as simple as the basic Freytag triangle, nor as arbitrarily complex as models of individual deep stories can get (which need only begin and end with a liminal passage).

We will adopt the Double Freytag triangle as our canonical example of the structure of a deep story.

The previous chapters have armed us with a compact vocabulary for describing the various phases of the subjective experience in the Double Freytag triangle, so let's take a tour.

4.4   A Canonical Deep Story

The Double Freytag triangle is a representation of a prototypical deep story in many domains. College, marriage, or a transformative experience at work (such as managing a product launch) can all take on this form.

The enact a deep story you must remain aware of the tempo of different phases, manage the emotions they evoke, and time various shifts in optimal ways.

4.4.1   The Liminal Passages

Characteristic tempo: stillness

The bookend liminal passages, as the only necessary features of a deep story in our model, are the most important elements of the Double Freytag triangle.

Recall from our discussion of archetypes and doctrines that these are the main persistent elements of all our mental models. During liminal passages, they occupy center stage. Between the initial and final liminal passages, doctrines and self-archetypes evolve.

Here is an example. Suppose you are young, idealistic and operate by the belief "I can do anything I set my mind to." You experience a deep story during the course of which, despite your best efforts, a random act of nature (such as an earthquake), destroys your work. You might extract a moral from the story: "outcomes are never certain." You might then integrate the moral into your doctrine and self-archetype in the form of a more mature and sophisticated life philosophy, "I can take on anything, give it my all, and walk away content even if I fail. What doesn't kill me only makes me stronger."

During liminal passages, we are caught in the situational emptiness between rich narrative contexts, so the only thing we *can* do is indulge in existential musing. The musing can be either pleasurable or painful, and you can engage in it in either self-aware or self-indulgent ways. It can be so painful that you might be tempted to avoid it by getting drunk, or so addictive that you might get trapped for an extended period, possibly the rest of your life, reluctant to move on to the next deep story. This is one way to understand narcissism: it is falling in love with yourself at a particular liminal passage to such an extent that you are unwilling to submit yourself to further transformative experiences.

What exactly happens between the before and after liminal passages?

4.4.2 Exploration

Characteristic tempo: volatile, dissipative

In a classic *Dennis the Menace* cartoon, we see Dennis looking for something. His buddy, Joey asks him what he is looking for, and Dennis replies, "I'll know when I find it." That open, random and purposeless mode of cognition is the characteristic feature of exploratory behaviors.

We've identified learning, in the most general sense, as the process of constructing a mental model from scratch. This process is open-ended and has no goals beyond hardwired biological ones. It is unsupervised, uncertain, unbounded, unstructured, and mostly unrewarding. In more familiar terms, there are no teachers, safety belts, syllabi, grades or prizes.

Given these characteristics, it should not be surprising that it is a very disorienting and stressful phase in a deep story. Things you don't know that you don't know (unknown-unknown beliefs) dominate the situation. This is the "blooming buzzing confusion" that William James speculated, newborns experience (recent research suggests that this phase is more complex than James thought [27]).

Fortunately we are born equipped with several *natural* behaviors that appear to have evolved specifically for such situations. They are the bootstrap routines of our brain. All such natural exploration behaviors are varieties of *random* behavior. This should not surprise you. When you know nothing, and have no goals, random behavior is the best you can come up with. Within random exploration there are variants such as "hide and watch," "play" and "poke with a stick." We will encounter these and other non-exploratory primitive behaviors in Chapter 5.

Within a deep story, raw information with which to build a mental model is accumulated through exploration. The inherent open-ended randomness of all exploratory behaviors leads to randomness in what is discovered. To a child, this is stimulating variety. To the impatient adult, this can seem like inefficient collection of irrelevant information. Exploration is a process that increases both the size and disorderliness – or entropy – of a developing mental model of a fundamentally new situation.

4.4.3   The Cheap Trick

Characteristic Tempo: Crescendo

The cheap trick is to a deep story what the formation of a theory of the situation is to an ordinary enactment. The operative words in my definition of narrative rationality are *compelling and elegant*. The cheap trick introduces this compelling elegance into the evolving story.

The anxiety and incoherence of exploration cannot increase indefinitely. Whether or not we have enough information to act effectively, the sheer cognitive stress of exploration makes us seek relief, even when it takes the form of safe play among children. Our minds demand relief, and this leads to the moment I call the cheap trick, when the trajectory of increasing dissonance and entropy is arrested and turned around.

The moment occurs when you recognize an *exploitable pattern* in the raw material you have collected in your exploration. The immediate consequence of this recognition is the drawing of a relevance boundary.

Things that conform to the pattern are deemed relevant and included in the mental model. Things that don't conform are excluded. The unexplained chaos in your head separates into exploitable meaning and ignorable noise.

It is a phenomenon that has many other names, ranging from the Aha! moment to Clausewitz's poetic *coup d'œil* (strike of the eye). So why have I chosen to use the term *cheap trick?*

The short answer[†] lies in the famous H. L. Mencken quip, "for every complex question, there is an answer that is simple, elegant and wrong."

In a deep story, the default, driving question is, "What's going on, and what should I do?" The cheap trick provides an answer to this question, in the form of a key organizing insight that motivates the action in the rest of the deep story. Every such insight is flawed, since it is based on excluding some part of reality as noise. This will eventually catch up with you, so the insight merely buys you a certain amount of time.

If your wrong answer happens also to be *elegant*, it will compactly explain the part of reality that you do include, and provide leverage. If this leverage can bring you rewards within the time you've bought, you've cheated nature: earned real rewards from fake truths, and fled with your ill-gotten gains before nature takes her revenge through un-intended consequences. The cheap trick is the insight that allows you to locally and temporarily trick nature into bestowing disproportionate rewards on you.

Timing is critical. If you seek relief from the dissonance and stress of the exploration phase too early, you may latch onto a cheap trick that buys you too little time, and whose elegance provides too little lever-age. You will make an ineffectual stab at exploiting the environment for rewards, and fail to escape the unintended consequences. This is what computer scientist Donald Knuth had in mind when he said, "premature optimization is the root of all evil." Waiting too long isn't a great idea either, since it is a route to death by perfectionism.

The battle between your impatience and your ability to endure dis-sonance creates the moment. By managing the battle, you can influence the timing of the cheap trick, but not entirely control it, since it is de-fined by the existence and recognition of an exploitable pattern. If you fail, you will fall victim to either expediency or perfectionism.

[†]There is a longer and more solid answer that relies on the theory of computational complexity[28] and empirical studies of solvability of theoretically difficult problems.[29, 30]

This is what turns decision-making into an art rather than a pure science. The cheap trick is the spotting of an exploitable pattern in the information you have gathered during your exploration. This information is necessarily local in time and space, and incomplete with respect to all the potentially relevant information. There is no known systematic method for triggering a cheap trick.[‡]

In narrative rationality, the cheap trick is a more fundamental mental event than goal setting. Where a calculative-rational decision-maker might consider fair costs and actual values of various desirable goals in a new environment, the narrative-rational decision-maker looks for steals and bargains.

This is one reason Napoleon is an iconic figure in narrative decision-making. In war, finding a way for a small force to defeat a large one is generally an impossible problem. By only picking battles where he could find a cheap trick, or *coup d'œil*, he managed to rack up a string of apparently impossible victories.

A very special cheap trick is worth noting. Every cheap trick is about drawing a relevance boundary in a set of information discovered during exploration. Theories of early childhood suggest that the first such boundary we draw is between our sense of self (or "I") and what lies outside. A baby recognizes the relevant pattern that some elements of sensory reality appear to be under its control.

This very first cheap trick creates the core of all future mental models, a model of ourselves, a self-archetype. The first self-archetype is merely the subset of James' blooming, buzzing confusion that we classify into our mind's "I." It is extraordinarily hard to redraw this boundary later in life. Hallucinogenic drugs, intense stress, sensory deprivation chambers, or the rigors of meditative practice are usually required.

4.4.4 Sense-Making

Characteristic Tempo: decrescendo with emotional relief

Once you have spotted the cheap trick – a pattern that you can exploit to extract rewards from an initially incomprehensible environment – you must reflect on and reorganize the discoveries you made during exploration. A cheap trick is not just an exploitable insight, it is an *or-*

[‡]And, I believe, there never will be, unless an unproven conjecture in computer science, known as $P \neq NP$, widely believed to be true, turns out to be false.

ganizing insight. It serves as a speck of dust around which a compact mental model can crystalize and grow, within the turbulent soup of data created by your exploration. It allows you to make *sense* of what you've learned.

Imagine hunting around a room for pieces of an unboxed jigsaw puzzle. As you discover more pieces you initially organize them in some uselessly arbitrary way (green versus blue pieces for instance). But at some point you find a piece containing a fishy eye, and that leads to an Aha! moment (the green is a fish, the blue is water). You are now able make a great deal more sense of what you are looking at.

This is why "most compelling and elegant story" is the guiding heuristic in narrative-rational decision-making. Once you find a cheap trick, you can organize what you know in a very compact way. This compression and compaction creates a mental model where the pieces fit together in a meaningful way and lend the model significant and coherent momentum, like a generally disturbed weather pattern coalescing into a tornado.

If you have some familiarity with the philosophy of science, this should remind you of Occam's razor: the idea that the simplest explanation of a set of facts is probably the correct one.[§]

To make sense of a complex, ambiguous and confusing set of facts, you should look for an organizing insight that dissolves the complexity and provides you a compelling and elegant way to look at your situation.

To be *compelling* your view must be comprehensive and provide you a way to organize as much as possible, from what you know.

To be *elegant*, the resulting mental model must be as compact as possible. In general, these models will be very local and unique to the immediate situation.

By contrast, any applicable mental models derived from Jomini-style thinking (for example the five-forces Porter model for competitive business strategy) will be general, and lack the leverage-providing core cheap trick.

You can also understand this compaction as a process of decreasing entropy. While exploration leaves you with a large and disorderly collection of observations of the world, the entropy-lowering capacity of

---

[§]"Simplest" can be objectively characterized using a forbidding mathematical idea known as Kolmogorov-Chaitin complexity, which the ambitious among you can tackle through Gregory Chaitin's relatively accessible popular treatment *Meta Math*.[31]

the cheap trick allows you to make sense of what you know.

Elegant and compelling, however, do not add up to real. Remember, you are only providing an elegant and compelling account of what you already know. Your mental model has not yet weathered an encounter with new realities. Until it does, it is indistinguishable from a delusion.

As George Box noted, "All models are wrong, some models are useful." Your elegant mental model *will* eventually be proved wrong to a greater or lesser degree. Whether it is useful depends on whether it can get your through the next phase of the deep story: the valley.

4.4.5   The Valley

Characteristic Tempo: steady, with slowing momentum and increasing depression

The valley is a phase of initially rapid, and then slowing momentum development, eventually followed by a return to increasing entropy. In the valley, you encounter diminishing returns from the organizing capacity of the cheap trick. As the leverage provided by the cheap trick is exhausted, the enactment requires increasing amounts of raw energy. You are sustained only by the belief that you are cheating nature on a grander scale, and that beyond the valley lies a reward of disproportionate magnitude.

The characteristic feature of the valley is decisive action without either reward or validation.¶ Your enactment is charging ahead in the dark.

If you were to come up with a new scientific theory that permits time travel, the valley would be the period during which you build your prototype time machine. Until you turn it on, your actions must be driven by your leap of faith, and you must endure thoughts of failure.

A particularly interesting kind of valley is one where little or no action is required; one that merely represents a waiting period between

---

¶I derive the term from Psalm 23:4, "Even though I walk through the valley of the shadow of death, I will fear no evil..." The idea is related to the idea of the "Dip" introduced by Seth Godin, and the "trough of disillusionment" in the diagram known as the "Hype Cycle," introduced by the analysts at Gartner. Be careful about drawing analogies too quickly though; remember, a deep story is a graph of entropy in a mental model over time. The Dip and Hype Cycle diagrams plot different things, and contain subtly different assumptions.

the adoption of an intention and the first action that irreversibly commits you to the course. Shakespeare again helps us out:

> Between the acting of a dreadful thing
> And the first motion, all the interim is
> Like a phantasma, or a hideous dream:
> The Genius and the mortal instruments
> Are then in council; and the state of man,
> Like to a little kingdom, suffers then
> The nature of an insurrection.

Julius Caesar, Act II, Scene 1.

Here Brutus is experiencing his personal valley in the deep story of the assassination of Caesar.

The valley is the longest and most difficult phase of a deep story, but curiously, it is hard to say anything about it.

It is not just non-fiction treatments that lack the words to describe it. Fiction has the same problems. In movies about underdogs winning sporting contests, this is the part that screenwriters skip over lightly, with the help of a montage set to inspirational music.

4.4.6 The Heavy Lift

Characteristic Tempo: a high-effort, low-coherence increase in momentum

We exit the valley with a massive effort of the will: the heavy lift. If you've ever pulled an all-nighter to finish a major project, or powered through, despite doubts, to launch the product of your efforts into the wider world, you've experienced the heavy lift.

It is easy to confuse the heavy lift with the effects of deadline pressure. A heavy lift is *always* required in a deep story, whether or not there is a deadline. In fact, an arbitrary deadline of no real-world consequence is often imposed to force the heavy lift, and the *do or die* character of the effort is what makes it a *dead*line. In any consequential situation, the decision to start a heavy lift is deliberate, not forced.

As the valley stretches on, providing no validation or reward, it becomes progressively harder to sustain a disciplined tempo and steady effort. Like all mental models, the mental model being created by the deep story is unstable. It starts to disintegrate in the absence of a continuous stream of reinforcement through situational feedback.

Therefore at some point a steady tempo becomes unsustainable and the heavy lift is triggered. Subconsciously, we realize that it makes more sense to use up all remaining reserves of emotional resilience and energy in one final burst of fierce effort resulting in a forced encounter with reality.

The timing of the heavy lift is driven by personality and temperament. You launch a heavy lift when you are unable to stand the valley any longer, and sense that you are approaching diminishing returns. Whether this timing is *optimal* depends on your estimate of the organizing capacity of the cheap trick. If you overestimate its capacity, you will stay in the valley too long, and your act of creative destruction will fall victim to the perils of perfectionism. If you emerge too early, you may find you have not extracted enough value from the cheap trick to earn the disproportionate returns you dreamed of.

Entropy increases during a heavy lift because in driving towards a forced outcome, you inevitably begin making expedient decisions out of exhaustion, which leads to imperfections and compromises. As writers like to remark, books are never finished, they are merely abandoned.

This acceptance of necessary expediency leads to the increasing doubt and anxiety characteristic of the last hours before the first significant encounter with reality: the separation event.

4.4.7 The Separation Event

Characteristic Tempo: Crescendo

The separation event is the moment when a significant proportion of the newly created mental model, along with its momentum, is externalized into the environment, as your act of creative destruction.

It is no longer entirely within your control. At this point, if you were to follow the externalized portion, you would be following the course of a social or material dialectic (see Chapter 3). This moment is commonly referred to as the moment of truth, since it typically either validates or invalidates the assumptions underlying the leap of faith made during the valley and heavy lift.

The separation event sometimes coincides with what, in calculative rationality, is considered the achievement of an objective. But more often, it does not. A PhD program ends with the formal award of a degree (the "objective") but the separation event is actually the dissertation defense. The objective in launching a new company is attained when it starts to repay the original investment, but the separation event is the product launch.

Besides being a major and irreversible encounter with reality, the separation event involves a parting of ways, and the metaphor of birth is not accidental. The externalized mental model is an organic entity, like a baby, and there is a psychological imperative that pushes us to let go when we judge that it can live outside our heads. There is relief, but there is also a true sadness, since we can no longer protect the product of our efforts.

For a deep story like *College* which is primarily about recreating yourself, the separation event is the first true test of your new identity, such as the submission of a résumé and transcript to a recruiter.

Through externalization, a part of the mental model created by a deep story becomes codified and embedded into a reality upon which others will place their own constructions, and imbue with their own meanings. Beyond the separation event, we can only choose to participate in the construction of shared mental models, through an external dialectic.

4.4.8   The Retrospective

Characteristic Tempo: Decrescendo and a mix of joy and sorrow

The retrospective is the phase during which the decision-maker, if he or she has survived the separation event, attempts to return to the beginning state undergoing as little subjective change as possible, and receiving only an objective, externalized reward. In this phase, the decision-maker's doctrine is also revised, to reflect the morals of the deep story just experienced. The deep story itself, as a memory, is cast into its final stable form, in a way that validates the revised doctrine. We attempt to put the deep story behind us and continue life with the most valuable doctrine we can craft, not necessarily the most truthful one.

Since we rewrite history to support this expedient doctrine, retrospectives can lead to delusions as easily as they can lead to wisdom. The fox in Aesop's fable chose to remember his story as "the grapes are sour," rather than admit that he wasn't clever enough to get the grapes.

That way he was able to avoid modifying his beliefs about his own cleverness.

The retrospective is not the same as a calculative-rational debrief or assessment that might follow the accomplishment of a calculative-rational objective. It is a psychological consequence of either success or failure that can take as long as a lifetime to run its course. For veterans of major wars, the retrospective can last a lifetime (the character of Walt Kowalski, played by Clint Eastwood in the film *Gran Torino*, is an example).

The extent to which we are able to return to a low-entropy liminal passage is determined by the extent to which we are able to make sense of the impact (or lack thereof) of our post-separation actions.

If the separation event is successful (a successful dissertation defense, a great opening weekend for a movie), the cheap trick that drove the deep story is validated, and we extend it to explain and organize the new realities it has weathered.

If the separation fails, we face an existential crisis. We must search for new meaning amongst the ashes, deny and isolate ourselves from the aftermath, or simply remain in a high entropy state, mourning over the corpse of our failed efforts.

A popular Bollywood song from the 1960s poignantly captures this state of mind:

> We sit, having draped in a shroud
> the corpses of our dreams
> and watch the theater of our fates[‖]

While all deep stories transform us, failed separation is particularly devastating when the primary outcome is the hoped-for transformation. We cannot easily love again what we have just attempted to destroy and recreate. When *College* as a deep story fails, (such as not finding rewarding work that validates and strengthens the hard-won adult social identity) young graduates may feel compelled to find themselves in other ways, perhaps through experiences such as backpacking on a different continent.

[‖]*Kafan se dhaank kar baithe; Hain hum sapnon ke laashon ko; Jo kismat ne dikhayaa, dekhte hai un tamaashon ko*, from *Sambandh*, 1969

Even when the separation event is successful, a full return to the low entropy state that marks the initial liminal passage is rarely possible. Even if the world hails the creation, the creator is always aware of the compromises and imperfections that marred the separation event, due to the exhaustion-driven timing of the heavy lift, and the tragedy of expedience inherent in the cheap trick.

With the benefit of hindsight, the stable memory of the deep story, as well as a revised doctrine and self-archetype take shape during the retrospective. Whether your reconstructed memories are delusions or critical histories depends entirely on your capacity for honest introspection.

As the retrospective tapers off, once again we enter a liminal passage, where the experienced tempo slows to a near-standstill.

4.5 Narrative Time

So far we've encountered two sorts of time: oscillatory and event-based.

Recall the David Landes quote at the beginning of Chapter 2. His question is a very intriguing one. Why after all, do we need to measure time using an oscillatory phenomenon?

As you may have guessed by my introducing the notion of entropy into our discussion, we are working towards a way to correct this unnatural state of affairs. We are going to start thinking of time in terms of a unidirectional phenomenon, entropy. It won't be even or continuous, but as we will see, those requirements are only critical for calculative rationality. Narrative rationality *necessitates* a bumpy, uneven ride.

Event time, which we encountered in Chapter 2, brings us a little closer to a unidirectional sense of time. It helps us understand time as a metaphoric container for things that happen which in turn gets structured by the things we actually put into it. Event time is more general because we can represent both periodicity and aperiodicity in the change that drives forward both the universe at large, and our individual lives. And we can represent entropy too: time as a container of events can get more or less cluttered as it stretches into the past and future. But the container itself is still even, uniform and bi-directional. There is no necessary distinction between the past and the future. To get to a true narrative sense of time, we will need to drop the container and keep the contents.

Together rhythms and events help us think of time as *material change*. And if you think about it, this is the only way we *can* think about time. If things weren't changing, it is not clear that we would experience a sense of time passing. The specific sort of change that you use *does* matter. Whether you prefer the natural flooding of the Nile, the phases of the moon, your own breathing or the "unnatural" rhythms within cesium atoms (atomic clocks), the choice of a temporal standard of comparison matters for narrative rationality. Every sort of narrative time is a *specific* pattern of change, with a characteristic texture of periodicity and eventfulness.

But we can say one more thing about time: it appears to go in only one direction, and the best way to think about the direction appears to be as the direction of increasing entropy. This is what physicists call the "thermodynamic arrow of time." The direction appears to be the same as what we call the "psychological arrow of time" (our subjective sense of past versus future).

What this curious coincidence means is a matter for philosophers, but for us, the arrow of time leads to an idea that I will call the Freytag staircase.

4.5.1   The Freytag Staircase

Entropy is part of what characterizes our unidirectional sense of time, and helps distinguish it from the platonic, reversible models of time assumed in calculative rationality. From the point of view of decision-making, the future and the past differ primarily in that the future is more disorderly and complicated. Just as water may flow upstream locally and temporarily in eddies, entropy can be lowered locally and temporarily. But the universe as a whole appears to be headed in one direction.

What happens when you view your life as a string of deep stories, with each successive liminal passage being, on average, a little higher than the previous one?

The difference between the initial and final liminal passages can be interpreted equally well as doctrinal growth, or decay. Since we are interpreting the vertical axis as entropy, we will adopt the latter interpretation. This upward path is the Freytag staircase, and its pattern of rise and fall is the fundamental tempo.

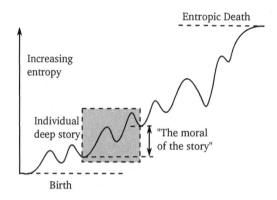

The Freytag Staircase

Occasionally, a deep story will end having lower entropy than when it started. A profound experience during your college years might lead to you figuring out your entire life, rather than just supplying an identity to serve you in your twenties.

But on average deep stories display a pattern of rising entropy, punctuated by liminal passages. Your personal life doctrine is a gradually growing set of irreducible axiomatic beliefs. As you accumulate transformative experiences, your doctrine starts to occupy increasing amounts of room in your head, limiting the capacity for open-ended thinking. Ironically, even the belief that one must be open-minded is doctrinal, and adopting the belief makes you less capable of living by it.

The Freytag Staircase is a structure. What can we say of its contents? As with individual deep stories, the definition of narrative rationality implies that the evolving doctrine might be delusional. The study of such delusional evolving doctrines is fascinating, and a great starting point is *The Redemptive Self* by narrative psychologist Dan McAdams,[32] in which he describes a particular pattern of life-doctrine delusions peculiar to successful Americans.

Whether we are self-aware or deluded each of us through introspection continually re-examines, prunes, consolidates and compresses our doctrine into a minimal set of self-evident and mutually independent beliefs. Introspection as a process is uncannily like trash compaction:

your head fills up with unrelated, irreducible beliefs.

It does not matter how thoughtful your doctrine is. The larger it gets, the more rigid and constrained your behavior becomes. This is why the word *doctrinaire* is synonymous with *rigid*. A growing doctrine is a path of increasing entropy. The Freytag staircase is a stairway to high-entropy heaven.

The Freytag staircase helps us think about cradle-to-grave life narratives. The rules for sketching one are a simple extension of the rules for sketching a deep story: any unbroken graph drawn between a minimum-entropy floor (babyhood) and a maximum-entropy ceiling (death) is a valid Freytag staircase. It can be smooth or jagged, and can take a long or short time to get to entropic death. In between it may very nearly kiss the ceiling and touch the floor, a metaphoric death-and-resurrection. Every peak is either the cheap trick or separation event of a deep story, or the climax of a simpler enactment. Every trough is either a liminal passage or valley within a deep story, or a context switch.

Try drawing a few Freytag staircases to get a sense of what you can visualize with this construct. For example, a very smooth trajectory might represent the idea of lifelong health based on the idea that good health is the slowest possible rate of dying. A trajectory with many small waves (high-tempo) is probably less philosophical and more active than one with a few big swings. A floor-to-ceiling "rebirth" swing standing between an extremely jagged before trajectory from an extremely smooth after trajectory, might represent the life of a mystic or prophet.

But even the most thoughtful among us cannot compress doctrines to nothing. Even if we occasionally manage to clear our heads completely and achieve Zen no-mind in a moment of enlightenment and spiritual rebirth, the state is unstable. Real life creeps back in, and the doctrine starts growing again. Even (and perhaps especially) if you attempt to hide out in a monastery and preserve that precious moment of rebirth for the rest of your life.

No matter how you draw your staircases and interpret them, they are valuable because they keep you grounded in the observation made at the beginning of this chapter: all our choices are among life stories that end with our individual deaths. You can visualize this by representing a significant decision as a fork in a Freytag staircase.

Ultimately, both staircases wind their way to the same entropic-death state. There is a sense of proportion and pensive comfort to be found in

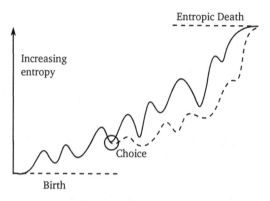

Visualizing decisions as choices
among Freytag staircases

this thought.

4.5.2 The Fundamental Tempo

The Freytag staircase is rhythmic. It has a fundamental tempo driven by the rise and fall of deep stories, aperiodic (not a stable, even oscillation, but noisy) and entropic (drifts in a specific direction). In other words, it is a subjective, narrative model of time, a narrative clock for an individual life.

This may seem like a weird idea. The Freytag staircase is easier to understand as a construct *based* on time – a graph of clock time versus entropy – than as a model of time itself. But imagine letting go the reassuring foundation of familiar clock time.

What happens if you remove the time and entropy axis lines, and even the surrounding space? Imagine that the Freytag staircase is actually a roller coaster that you are riding with your eyes closed. You have no external reference points; your only sense of time passing is based on how you *feel* about what you are doing. Sometimes you are climbing, sometimes you are moving along level ground and sometimes you are going downhill. Time seems to flow faster or slower based on your emotions.**

** If you have had some exposure to differential geometry, you will recognize the idea

The legend of Sisyphus, condemned by vengeful gods to roll a rock up a hill, only to see it roll down again for all eternity, is a particularly fertile fable about narrative time, which helps explains why Camus was able to extract so much insight from it, in *The Myth of Sisyphus*.[33]

When you describe the temporal context of your life in this rich way, rather than just marking off a metronomic sequence of ticks on a straight line, you catalyze narrative rationality. Narrative rationality deals with decisions against the backdrop of an appropriate narrative time, measured by an appropriate narrative clock. As Lenin once remarked, "There are decades where nothing happens; and there are weeks where decades happen."

By contrast, calculative rationality is based on the fiction of an abstract time that stretches evenly and featurelessly into the past and future. This is why regularity, stability and precision matter.

Calculative rational decision-making finesses situatedness by working only with highly controlled clocks. When you start this way, seductive general theories of decision-making take center stage, and local conditions are forgotten, minimized or dismissed as "details." To achieve this apparently legible and clean state of affairs, calculative rational models of decisions seek the most precise, featureless and stable clocks available.

4.5.3 Personal Narrative Time

Narrative time makes some things harder. A single narrative time, unaided, cannot achieve precise coordination across multiple narratives. A clock based on "when the cows come home" cannot easily interact with a clock based on "when the Nile floods."

The original impetus for the development of modern time zones was precisely such a coordination problem: getting trains to run on time. The anarchy of narrative times across the breadth of America posed a serious problem for the first railroads. The problems were solved, but the cost was the homogenization of time wherever railroads went. The American Midwest still runs on a more leisurely narrative time than the coasts, but the differences we see today are impoverished remnants of the diverse temporal ecosystem that existed two hundred years ago.

---

of intrinsic coordinate systems in this metaphor.

But narrative time also makes some things much easier, especially at the level of individuals and small-scale interpersonal interactions.

If I were to approach you with a proposal for a collaboration after you have read this book, that interested you, but you were unable to commit at that moment, you might react in one of two ways.

The calculative rational way would be to say, "I am overbooked right now, but I'd like to talk next year, maybe in January."

The narrative rational way would be to say, "I am in the valley of a deep story right now, how about I connect when I get to my next liminal passage?" Yes that sounds theatrical, and I doubt my language will catch on, but it helps to *think* in terms of narrative time.

Earlier we talked about the idea of calendar art (Section 2.7.9).

For you as an individual, time is not a simple label like "time (seconds)" on the axis of graphs. It has its own texture. Take a look at your calendar in an email program like Outlook. You see a tapestry of events woven together with interval logic. There are rhythms and arrhythmic elements. There are chaotic weeks and orderly weeks. Imagine coloring each event either red or blue, to indicate the emotional content. Imagine shading it lightly or darkly, depending on the momentum the events contain. Step back. What you have looks like a colorful river. If you imagine a calendar of your life stretching from birth to death, the river will sometimes contain rapids, sometimes flow gently.

Your calendar, of course, is merely the socially situated side of your personal narrative clock. Chronobiology is the other face of this clock. We will not attempt to cover this vast subject, but it is worth reading one of the many excellent popular treatments.[34, 35]

4.6   Narrative Models and Particularism

Even in the best hands, the Freytag triangle, the monomyth and their derivatives as templates for storytelling can lead to formulaic and clichéd results. As time standards, they are an improvement on calculative-rational clocks, but are still relatively impoverished with respect to the diversity of narrative times in our world.

The sophisticated novel reader or movie-goer gradually loses his ability to identify with, and vicariously experience, the patterns of energy and emotion experienced by the protagonists of formulaic stories.

You can tire of riding the same roller coaster repeatedly. The fact that deliberately crafted stories, as opposed to real-world events, contain cues that help us anticipate peaks and troughs, ends up blunting our responses to them. When individuals and cultures age, storytellers must work harder to trigger experiences of the same intensity.

As a theory of art and literature therefore, any structure like the Double Freytag triangle contains the seeds of its own destruction. Modern literary theorists, in fact, do not take such models seriously, and favor an approach called particularism, the idea that every story must be understood on its terms, without reference to general models or universal features.

Campbell's monomyth, for instance, is criticized for its highly speculative and overwrought metaphoric character. The heavy influence of Freudian and Jungian thought on the model, in particular, leads to some very valid criticisms of the specifics.

But at a coarse level, models such as the monomyth undeniably capture some universal characteristics of storytelling and experienced narrative time that resonate with us.

The Double Freytag and Freytag staircase models are partly an attempt to synthesize particularist and generic views of time. I offer them only as partial models of prototypical narrative patterns. The patterns usefully illuminate the ideas of dissonance and entropy in mental models. Apply them with taste and discretion.

4.7 Mortality and Rationality

Risk, learning and information are central to calculative rationality. The idea of mortality necessarily looms large over these topics, but is at best a marginal concern in calculative rationality, restricted to those who deal with problems that explicitly deal with mortality, such as insurance.[††] By contrast, in narrative rationality, mortality is the central concern.

The relationship among risk, learning and information seems deceptively simple: every decision is based on what you know (information), and risk assessments associated with what you *don't* know. *Learning* helps you increase usable information and lower risk.

[††]In other domains such as investment, even professionals routinely forget about mortality effects (such as investors losing all their wealth and exiting the market, leading to the well-known survivorship bias effect).

Calculative rationality focuses on risks (specifically, risks that can be modeled *a priori*), learning and information primarily because you can bring a great deal of very sophisticated mathematics to bear.

Empirical probabilities derived from historical data can be used to estimate the likelihood of different futures. Sophisticated discounting models can roll up the value of future payoffs into the net present value of likely options. And finally, a host of automated learning mechanisms can crunch raw data about outcomes into information about future expectations and even automated decisions.

So why is this picture of comprehensibility deceptive? The reason is that despite the awe-inspiring mathematical machinery, such models are still based on an assumption that runs through almost the entire discipline of statistical decision-making like a giant fault-line: the future is like the past.

The models are only as good as the instincts of the model builders in tuning the parameters, the accuracy of the model with respect to the past, and your level of faith in future-is-like-the-past assumptions in a given situation. The models can be viewed as a way of systematically leveraging or amplifying intuition.

But intuition amplified through calculative-rational models, while necessary, is not sufficient for true risk management.

4.7.1 Open and Closed Worlds

Future-is-like-the-past assumptions are a symptom of what are technically known as *closed-world* models: models in which pre-defined things exist, and only events that have been modeled can happen.

The real world, by contrast, is an *open world* that requires that we constantly update our mental models to accommodate phenomena that we *haven't* encountered before. The open world is a world that includes what Donald Rumsfeld called "unknown unknowns" and Nicholas Nassim Taleb calls "black swans" (rare, highly consequential events). This *necessarily* requires accommodation of periods of high entropy in mental models, while fundamentally new and unexpected information is being incorporated.

Informal human mental models can comprehend open worlds because they can contain things that haven't been understood: a high entropy state. Through the modulating effects of emotion on cognition, we seem capable of computing with "clear as mud, but covers the ground" high-entropy mental models. Artificial methods for making good use of poorly understood, high-entropy information, are currently in their infancy.

In open worlds, understanding is a live, ongoing battle to lower the entropy of the information in active mental models. And this ongoing understanding of the world, as we saw in our discussion of the Freytag staircase, weakens over time for any decision-maker with finite capabilities. As we age, we become more doctrinaire and less capable of open-world learning. Narrative-rational decision-makers necessarily age over time.

Calculative-rational decision-makers though, need not age. The study of calculative rationality has focused on different sorts of archetypal decision-makers over the last few centuries: perfectly logical decision-makers, bounded-rational decision makers and most recently, predictably irrational decision-makers. But all members of the calculative-rationality family tree are immortal. They are ageless bundles of logical reasoning capabilities, processing limits and systematic biases. They may have memories and histories, but they do not grow old.

In contrast, narrative rationality starts with *mortal* decision-makers. Mortality is the central fact about them. Decision-makers can die in two ways: accidentally, through the impact of unknown-unknowns, or through the accumulation of entropy, as the open world catches up with them. Thinkers such as Taleb have eloquently and elegantly considered the former, but for us, it is the second kind of mortality that is interesting.

4.7.2   Death by Entropy

Narrative-rational decision makers are mortal agents who have a fixed capacity for absorbing open-world information and battling entropy, before they succumb. They exhibit *entropic aging*. They climb the Freytag staircase and die. This idea of mortality of course, is philosophical rather than literal. I like to refer to this philosophy as thermodynamic theology.

You do not need to understand the laws of thermodynamics at a technical level to appreciate the core tenets of thermodynamic theology. For our needs, this irreverent (and surprisingly accurate) folk version of the laws is actually more appropriate. The three laws of thermodynamics are:

1. You cannot win

2. You cannot break even

3. You cannot quit the game

Some people add a zeroth law: *you must play the game.* There is no point in lamenting, like Kurt Vonnegut, "I didn't ask to be born." The calculative-rational rule of thumb offered by Stephen Covey, "win-win or no deal," does not apply to life. You cannot opt out except through suicide.

Complexity theorist Stuart Kauffman proposed a fourth law that can be stated as *the game keeps getting more complicated, and there are always more different ways to play.*[36]

Thermodynamic theology, as captured by these laws, is ultimately at the heart of narrative rationality. These laws, coupled with constructs like the Freytag staircase and narrative time, help us operationalize the idea offered at the beginning of this chapter: all our choices are among life stories that end with our individual deaths.

Which brings us to our skill for this chapter: Tetris thinking.

4.8   Skill: Tetris Thinking

When they get to subjects like risk, books on decision-making traditionally introduce analogies between life and games. Chess, Go, poker, casino games and football are the usual suspects. I prefer an unlikely candidate: Tetris.

Like other games Tetris is a closed world, but it models the primary processes in open-world risk management – dealing with increasing entropy and consciously choosing your path to death – very well. Playing Tetris helps you hone entropic decision-making skills.

If you have never played Tetris, I suggest you do so now before reading further (several free versions are available online).

The purpose of the game is to position and stack blocks of various shapes that fall from the ceiling to the floor of a tall, rectangular gameplay screen. Compact, filled rows at the bottom disappear by sinking through the floor, while poorly packed rows with holes persist. Over short periods, the height of the stack can go up or down, but over longer periods, it inevitably rises, and you have a gradually shrinking amount of vertical space above the stack to maneuver new blocks into place.

There is a vicious cycle: If your current decision is poor, your next decision becomes harder, since bad decisions raise the stack height, leaving you with fewer options and less time to make the next decision.

Eventually, the stack hits the ceiling, and you die. As the stack rises and falls, the evolution of the stack height over time traces out a Freytag staircase.

Chapter 5

Universal Tactics

> If an important decision is to be made [the Persians] discuss the question when they are drunk, and the following day the master of the house... submits their decision for reconsideration when they are sober. If they still approve it, it is adopted; if not, it is abandoned. Conversely, any decision they make when they are sober is reconsidered afterwards when they are drunk.
>
> – Herodotus

Universal tactics are primitive concepts of action, such as *push* and *pull*, that structure our behavior in all domains. They arise from a few universally experienced domains, such as space, time and matter. Whether you are in a battlefield, in a library solving a mathematics problem with pen and paper, playing music or doing chemistry experiments, your understanding of these universal domains shapes your behavior, via metaphor.

These metaphors arise from two features shared with our ape cousins that shape our experience of the world: opposable thumbs and rich, binocular, color vision. We are defined by our hands and eyes the way dogs are defined by their noses and teeth.

Tactile manipulation has become our metaphor for all action, even though as social creatures, our tongues are at least as important a mode of action. So we *grapple* with difficult problems and deliver *touching* eulogies.

Similarly, we have been shaped by our eyes to such an extent that *seeing* has become a metaphor for all perception. When you *hear* somebody make a *point* (a visual concept) in a conversation, you are more likely to respond, "I see" than "I hear you."

We routinely underestimate, however, the degree to which our hands are involved in perception. It is our hands that hold up and rotate small objects for inspection, and in the dark, help us see through groping. It is our hands that clear away clutter so we can see what lies beneath. It is our hands that shuffle through papers, allowing our eyes to scan as we search for a particular document. But while our eyes cannot do much, our hands can see independently of our eyes. For the blind, the hands literally take over for the eyes, through Braille.

Our model of universal tactics rests, ultimately, on this visual-tactile engagement of the world.

For example, a commander on the battlefield *separates* enemy forces just as a mathematician *separates* an algebraic expression into two simpler parts. Both are examples of a "divide and conquer" primitive, universal tactic. The basic intuition in both cases is drawn from the experience of manually separating a set of objects in space. In visual-tactile terms, we see an imaginary line through a set of objects, and sweep the objects away from the line using hands.

My notion of universal tactics is derived from the theory of conceptual metaphor introduced by George Lakoff and Mark Johnson in their 1980 classic, *Metaphors We Live By*.[37] Conceptual metaphors allow us to map our literal sense of an idea such as *push* to other domains (such as *push him to try harder* or *we are planning a major marketing push around Christmas*).

We will organize our views of human behavior around this fundamental eyes-and-hands metaphor for perception and action, emphasizing in particular the dominant role of hands in perception when visibility is poor.

5.1   Natural Behaviors

In extremely new situations, such as infanthood or the exploration phase of a deep story (Chapter 4), mental models are just beginning to form and are very weak. The only reliable core elements are the doctrine and self-archetype carried over from previous situations. When such weak mental models encounter novel environments, *natural* behaviors are produced. By contrast, familiar situations, such as driving a car, trigger learned behaviors that are part of richly evolved mental models. We will call the latter *artificial* behaviors.

Natural behaviors, such as aversion, are rudimentary, hard-wired and have evolved primarily to handle unknown risks and open-ended learning. They help bootstrap mental models. Artificial behaviors are generally much more complex, learned rather than inborn, and help manage known risks and close-ended learning within rich mental models.*

We briefly touched on risk, information and learning in the last chapter, and noted the fundamental relationship among them: learning helps us increase our store of usable information and lower risk.

In most situations, you can learn a lot faster by doing than by watching, but unfortunately, action also exposes you to more near-term risk than watching. This is because most environments, even dangerously unstable ones, are relatively quiescent unless disturbed. They do not reveal much under normal circumstances. You typically have to *do* something in order to provoke a reaction from the environment. Such provocation reveals useful information that can drive an enactment forward.

The classic example of this phenomenon is language acquisition. When a child points to a cow and says "doggie," and is corrected by a parent, a concept is learned and a hypothesis is refined. Progress is much faster than with pure listening, but at the cost of mistakes made along the way.

This creates a bootstrapping conundrum: how can you act meaningfully when you don't know much about the situation, and do not have relevant, well-developed mental models?

The answer, as we saw briefly in Chapter 4, is that you cannot. You must act meaninglessly, or in other words, *randomly*. Unstructured learning behaviors are random to a lesser or greater extent depending on the maturity of the active mental models.† In childhood environ-

---

*A third category of behavior, non-trivial hard-wired behavior, is not very important in humans. Insects such as bees exhibit many such behaviors, such as hive-building and caring for their young. Humans have to learn equivalent behaviors. This dominance of software over firmware, the effect of an adaptation called neoteny, makes us both more vulnerable than other species' young during our extended childhood, and more adaptable to new circumstances as adults.

†This insight, codified into a formal idea called the "persistency of excitation," is at the heart of many powerful learning and adaption techniques in control theory.

ments, parents create a level of safety, so random exploratory behaviors can be quite unrestrained. This creates the most basic natural behavior: "play." When children perceive risk, they naturally exhibit more cautious kinds of random, exploratory behaviors: "hide and watch," (where the randomness lies in how attention is directed) or "touch gingerly."

Adults cannot assume that the environment is safe, so adult natural behaviors exhibit a wider range of calibrated amounts of risk management. As with children, *hide and watch* is our reaction to dangerous-seeming and volatile situations (gather information passively while remaining out of sight). Colloquialisms such as "dip a toe in the water," "poke with a stick," "stir the pot" and "stir up a hornet's nest" are increasingly bolder modes of risk-managed exploration. The safer we feel, the more actively and aggressively we explore.

Note two features of natural exploratory behaviors: they are fundamentally iterative (the actions are designed to produce feedback, which triggers further actions), and they involve a certain amount of anxiety, which implies stress. Exploratory behaviors are naturally correlated with high-energy defensive behaviors (fight-or-flight) as a high-probability follow-on, and therefore involve anticipatory stress.

There is actually an idea in psychology that makes the connection even stronger: the Yerkes-Dodson law. This law posits that the ability to learn *requires* the right level of arousal. Too little, and you get boredom and disengagement. Too much, and you will react too quickly with *hide and watch* (or even *fight or flight*, the best-known natural behavior) to situations that might yield more learning with bolder exploration.

Natural behaviors therefore also have a natural unstable tempo driven by the rapidly changing level of caution in the iterative probing of the environment. A child can go from a very slow *hide and watch* tempo to a very high *play* tempo very rapidly. Except in the case of safe play environments, emotions are generally on the stressful, anxious end of the spectrum, and the energy level is poised unstably between the energetic bursting of fight-or-flight and the tense stillness of hide-and-watch.

While most natural behaviors are either literally or metaphorically tactile, *hide and watch,* the most cautious member of the group, relies on literal or metaphoric vision. We may move from one hiding place to another, in search of the best vantage point, but we do not actively influence the situation in order to learn faster. It is still random though, since we scan for structure in the situation rather than seeking to discover specific answers in focused ways.

In hide-and-watch situations, we prefer conditions that help us learn fast, while remaining safe. In studies that involved showing subjects a variety of photographs and documenting the stress levels induced, Rachel and Stephen Kaplan[38] discovered that humans are most comfortable in complex but legible environments.

As a rough approximation, for example, we tend to prefer lightly wooded environments over either dense, impenetrable forest cover or barren deserts. That particular balance between being able to see and avoid being seen was perhaps the right one for our species' needs in our early evolutionary environment. Similar principles apply to urban environments. We will examine the idea of complex-but-legible more deeply in Chapter 6.

Natural behaviors are generally either trivial or unpredictable. Artificial behaviors, on the other hand, can be extraordinarily complex. For instance, driving is estimated to involve around 1500 distinct subskills.[2] And that's just *one* domain of human experience. From cooking to programming to boarding an airplane, we deal with dozens of domains of comparable complexity. Social reality presents us with an extremely diverse and complex parade of situations, each of which calls for its own specialized behavioral vocabulary.

Complex artificial behaviors have their own evolved, domain-specific tactical vocabularies of action. These are born, mature, and die just like human languages. At one time, knowledge of sword-fighting was commonplace. Today it is rare; that particular behavioral vocabulary has been displaced by the tactical vocabulary that governs shooting.

But through the rise and fall of specific tactical vocabularies, an underlying universal vocabulary of tactics persists. This vocabulary rests on our capacity for thinking in terms of conceptual metaphor. The existence of this vocabulary is the reason why a game like chess is as valuable to modern military officers who must work with air power and missiles, as it was for military officers a thousand years ago who had to work with cavalry and archers.

5.2 Conceptual Metaphor

Like every language, the language of enacted behavior has a syntax and semantics. Syntax is described by grammar. Not high-school "don't end a sentence with a preposition" grammar, but a more fundamental sort known as a *Universal Grammar,* introduced by Noam Chomsky, that applies to anything that looks like a language.

Chomsky's idea of grammar applies not just to human language, but to such things as the structure of DNA, the behavior of robots, the structure of mathematical proofs, the formation flight of birds, and the design of computers. Fascinating though they are, we won't say much about syntax and abstract grammars.

We are more interested in semantics, the study of meaning. Specifically we are interested in an approach to meaning based on the idea of conceptual metaphor, due to George Lakoff and Mark Johnson.

A conceptual metaphor is a systematic structuring of meaning in one domain in terms of our understanding of another domain.

Here is a simple example: "He gathered his thoughts." Thoughts, literally, are patterns of neural firings. You cannot "gather" them. But the phrase structures our understanding of thought in terms of our understanding of tactile manipulation of a collection of solid objects.

Unlike isolated figurative metaphors such as "he was a lion on the battlefield," the expression "gathered his thoughts" is part of a more systematic mapping: "his thoughts were all over the place," "one thought stood out," "let's put those two thoughts together," "let's weigh the pros and cons."

Using the Lakoff and Johnson nomenclature, we would refer to this particular conceptual metaphor with the capitalized phrase THOUGHTS ARE THINGS.

Figurative metaphors are isolated embellishments within human languages. By contrast, conceptual metaphors are pre-linguistic raw material for the construction of meaning within mental models and enacted behaviors. Our use of Newton's laws and physics to understand the behavior of mental models is an extended application of the THOUGHTS ARE THINGS conceptual metaphor.

Conceptual metaphors lend meaning not just to the things we say with language, but to our perceptions of the world and to the behaviors we enact.

Here's an example of an observable behavior that can be understood through a conceptual metaphor: PEN-AND-PAPER MATH IS MECHANICAL ASSEMBLY. When you "group like terms" or "move x to the left-hand side of the equation" in the process of solving an algebra prob-

lem, your literal, observable behavior is rewriting lines of symbols repeatedly, making small changes each time. Yet, you primarily think in
terms of grouping, separating, combining, moving, plugging in, crank-
ing through to the answer, and so on, not rewriting. It is revealing (and
somewhat ironic) that computer scientists who build AI systems for domains like mathematical reasoning *do* think in terms of "rewrite rules,"
even though *their* literal domain is not pen and paper, but bits being
turned on and off in silicon chips.

We structure our thinking in any domain using conceptual metaphors
that map thoughts to a handful of universal visual-tactile domains, such
as dealing with solid objects and fluids, or inhabiting space and time.
This is one reason games like Go and chess are able to abstract away
from the details of the battlefield situations they model in such extreme
ways.

From these universal domains (universal in the sense that everybody
experiences them) we get our basic vocabulary of universal tactics. Later
in this chapter we explore how our everyday notion of the unqualified
word *tactic* relates to universal tactics.

Universal tactics combine to form basic fragments of enactments
which we will call *decision patterns*, which are analogous to sentences.
These in turn combine to form complete enactments. Since the roots of
this vocabulary lie in biology and our evolutionary history, you should
not expect clean edges. The language of action is just as messy as the
language of thought (based on mental models constructed out of beliefs,
desires and intentions) that we encountered in Chapter 3.

5.3   The Language of Behavior

Here is a quick look at the three most important classes of universal
tactics: those arising from spatial, material and patterning conceptual
metaphors. These are so fundamental that nearly all board games organize game-play in spatial and material terms, and involve pattern
metaphors.

To spot conceptual metaphors at work in real-world domains, it is
useful to develop an ear for metaphors in action-language. In the following inventory, I will list common examples in English that should
help you develop your sensitivity.

5.3.1 Spatial Tactics

Spatial concepts such as location, extent, orientation, distance, inside, outside, boundary, and overlap are perhaps the most important sources of universal tactics. When we map a domain to our understanding of space, we end up with a set of universal tactics based on fundamental spatial concepts. These include: moving towards, moving away from, inclusion, exclusion, containment, surrounding, touching, connecting, disconnecting and overlapping. Here are some simple examples:

- "His area of authority overlaps mine; we should define clearer boundaries." (influence over people as a region in space).

- "Let's propose 'bridging the gap between theory and practice' as the theme for the conference" (different aspects of work in a technical field as separated regions in space).

- "We need to outflank the enemy unit" (defining a boundary, controlling flow along and outside it, and confining the adversary's actions within it).

5.3.2 Material Tactics

Solid objects anchor our experience of concepts such as mass, roughness, friction, compressibility, extensibility, momentum, inertia and gravity. They also anchor our experience of counting and numbers. Universal tactics based on these include: resistance, yielding, dodging, dragging, attrition, rebounding and steering. Here are some examples:

- "I dodged the question," "I need to resist those conclusions," "he wore down the witness with his questioning." (framing a conversation as a game played with masses and forces).

- "We should lighten the burden of entitlements that is weighing down our company," "Let's spread the risk," "We are bundling different asset classes into a new type of security" (financial management moves viewed as load balancing, aggregation and disaggregation of solid objects).

- An interesting example from the battlefield is the hammer and anvil tactic from the days of cavalry, which involved trapping an enemy between a static or slowly advancing infantry line (the anvil, a solid object with a lot of inertia) and a cavalry unit at the rear (the hammer), moving more rapidly than the infantry.

5.3.3 Patterning Tactics

Through the order they can create in uniform space, solid objects also give rise to tactics that involve ordering, arrangement and pattern creation. Even though patterns seem conceptually derivative, our visual processing systems handle elements of pattern recognition (such as edge detection) at such low levels of brain function that it is useful to treat them as a fundamental source domain for universal tactics. Organizations are a particularly rich domain when it comes to patterning tactics.

- "He's next in line for promotion," (ordering organizational rewards using a linear pattern metaphor; employees are not literally standing in a queue waiting to be promoted)

- "He surrounds himself with yes-men" (encouragement of sycophancy as a ring around a center)

- "We need to add a competency area around statistics to our matrix model" (viewing an organization as an array of people-objects)

- "It's hard to get promoted around here; we have a flat organization with very little room at the top" (organizations as pyramids)

- "We need rotations for new employees so they can develop connections" (organizational divisions arranged in a circle; interpersonal relationships as a web)

- "Those guys in accounting are a toxic silo," (organizations as impervious cylinders)

- "Our industry is vertically integrated," (supplier-buyer relationships viewed as columns with respect to a "horizontal" market and supplier-supplier competitive relationships)

These spatial ordering metaphors for organizations can be traced to military precedents within which they were more literally evident. Ancient battlefields provide a particularly rich set of examples of patterning tactics. The Greek phalanx, a solid rectangular mass, is an example.

Textures can be viewed as a special class of patterns. The development of pike-based defenses against cavalry and later, barbed wire against infantry, provide excellent examples of texture-based tactics. The famous pike square used by the Swiss in the fifteenth century effectively created a large human porcupine to defend infantry against cavalry.

The development of firearms, and later, air power, created a more abstract, three-dimensional battlefield where patterning tactics have become more subtle, but military tactics are still created and expressed using the language of patterns.

5.3.4   Other Source Domains

Other important classes of universal tactics arise from our experience of fluids ("I am trying to make a *watertight* argument" or "the *surge* in Iraq is going well."), gases ("can you *expand* on that?") and mechanical assembly ("you need to *construct* a better argument").

At a more speculative level, universal tactics of discrimination, taxonomy and classification possibly arise from our experience of variety in nature (the "animal, vegetable and mineral" triad). Perception of colors, shapes, sizes, textures, geometric similarities and symmetries drives our behavior when we are faced with overwhelming variety. Solving a jigsaw puzzle or traveling in a foreign culture triggers these universal tactics.

How these other source domains derive from the more fundamental ones is an interesting question, but we do not have room here to exhaustively inventory all the primitive universal tactics and their respective source metaphors. Fortunately you already have a strong primitive vocabulary, so let's leapfrog to sentences.

5.4   Basic Decision Patterns

Primitive tactics combine to produce basic decision patterns. These can broadly be classified into four groups: opportunistic, deliberative, reactive and procedural.

Within our language analogy, these would be simple sentence types. Just as most of language is constructed out of four basic types of sentences (declarative, imperative, interrogative and exclamatory), complex behaviors are built out of the four simple patterns.

Individual primitive universal tactics rarely constitute meaningful behavior in isolation, just as single words rarely constitute meaningful sentences. Decision patterns are little nuggets of narrative meaning constructed through the coordinated enactment of several universal tactics. They cannot stand alone, but have the same rise-fall structure as full-fledged enactments, and take an ongoing enactment and the corresponding theory of the situation from one stable and meaningful state to another. If a decision pattern is interrupted, you must either try to go back to the original state or improvise, since the intermediate states, like incomplete sentences, do not add meaning.

For entertaining illustrations of this principle, watch for the moments in sitcoms when a character tries to recover from a conversational slip by steering an underway sentence towards a different meaning, or by turning it into a sneeze.

This rise-fall structure of a basic decision pattern is created by the temporary instability of both the world and the active theory of the situation as we act. Think of them as the equivalent of individual breaths in decision-making.

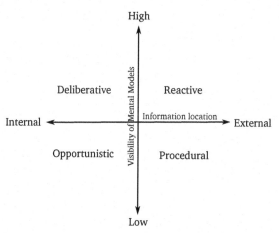

The four basic decision patterns

The distinctions among the four classes of basic decision patterns are not arbitrary. They are based on the distribution and visibility of situational information. Information either originates in the decision-maker's head or in the environment, and we either consciously recognize or are oblivious to the influence it has on our behavior.

A 2x2 matrix is *de rigueur* in a book like this one, and these distinctions supply the logic for ours.

Let's visit the quadrants in turn.

5.4.1 Reactive

Reactive patterns comprise a sequence of universal tactics that are consciously selected from among those suggested by the situation itself. They are almost entirely devoted to managing time, energy and momentum. The advantage of reactive patterns, of course, is that they can be enacted very quickly.

An example is looking for an opening to start a conversation with somebody you want to meet at a party. Reflect on how you actually do this: you might try to keep your target within view and reasonably close physically. A lull in the conversation, signaled by a brief silence or physical loosening of his or her current group, allows to break in with a comment. Or you wait for a moment of transition, when your target is moving from one group to another, or going up to the bar. When you spot your opening, you excuse yourself from your current group and move quickly to say, "Hello! I've been waiting to meet you."

If it is a large party with lots of people moving among groups and between rooms, with many people trying to intercept others (especially social magnets like beautiful, powerful or famous people), you might move from group to group or even room to room, and attempt to time your visits to the bar in synchronization with your target.

Here is how you know this is a reactive pattern: your decision pattern can be described by an independent observer with just one word – *following* (or *stalking*, if you are being particularly obsessive and persistent). Your complex behavior can be described very compactly with reference to your target's complex behavior. You have injected a fair amount of energy, but almost no new information into the situation. The information is also visible to you, since you are consciously using the tactic.

Our capacity for reactive behaviors is strongly related to our capacity for mimicry and imitation (which appears to be neuroanatomically fundamental, since we possess specialized mirror neurons that govern many basic reactive behaviors).

Examples:

- *Selection*: Saying "I am with you" in response to "Are you with me or not?" Tactics used: picking an object out of many. Source domain for metaphor: manipulating solid objects.

- *Reflection*: "You're stupid,"... "No, YOU'RE stupid!" Tactics used: catch, turn-around and return. Source domain for metaphor: manipulating solid objects.

- *Addition and removal*: "Yes, I agree, and to build on that..." or "I agree, but on the other hand, I'd like to push back a little..." Tactics used: attrition, accretion, compaction, pushing. Source domain for metaphor: dealing with solids.

- *Avoidance*: Pretending not to hear something in a conversation. Crossing the street to avoid meeting someone. Tactics used: looking away, moving away. Source domain for metaphor: space.

- *Negation*: "Let's go for Chinese food," "I'll go for anything except Chinese food." Tactics used: separation, aggregation, inclusion, exclusion. Source domain for metaphor: space and solid objects.

- *Joining*: Saying "me too" or making a voting decision. Tactics used: joining and subordinating identity to a larger entity. Source domain for metaphor: dealing with liquids or fine-grained solids.

- *Scan-to-task*: placing each of several unassigned resources "into" roles, such as in creating two teams at a pickup game. Tactics used: ordering, sequencing, matching and inserting in a container. Source domain for metaphor: patterning.

5.4.2 Deliberative

Deliberative patterns are those where the enacted sequence of universal tactics arises from prediction, inference and *a priori* computation, rather than mimicry. Note that deliberative patterns are not necessarily *deterministic* patterns. You can deliberately decide to act randomly for instance (a common choice in adversarial settings).

In a deliberative pattern, you inject new information into the situation. Your behavior cannot be entirely predicted or described with reference to something in the immediate environment. Since you must manufacture new information from old information to drive deliberative behavior, you generally need time to prepare. This is why we are impressed by people who can think on their feet very effectively. They are demonstrating a capacity for real-time deliberation.

Generating deliberative behavior requires computing with mental models. You simulate a few possible worlds consistent with the current state of your enactment. Whichever simulation gathers sufficient momentum to take control of your behavior first, generally wins.

During this race, many possible worlds are simulated in parallel, interacting and competing with each other for control. Most of the time, the momentum race solves the problem of which decision pattern to enact, with no conscious weighing of costs and rewards. This is because simulations cause a weak pre-experience of associated energy and emotion patterns, creating anticipatory attraction/aversion responses. We weigh options during the process of coming up with them.

This simulation runs a few steps ahead of reality, and relies on a logic that models the dynamics of the situation using the source metaphor. If you are structuring your enactment with a solid-objects metaphor, your logic will derive from the dynamics of solid objects. If your source metaphor is space, your logic will derive from the logic of time and movement. If your source metaphor is liquids, your logic will derive from the laws of flow. Precise argumentation is simulation that derives from our experience of mechanical assembly and construction, while looser forms derive from our experience of the dynamics of lumpy objects and liquids or gases (the prolific Edward De Bono distinguished between *rock* logic and *water* logic).

Deliberative patterns can involve initiation of reactive ones. For examples, you might carefully think through your elevator pitch for a business idea before a conference, where you hope to meet an investor. This would involve a good deal of simulation and selection. But the last step in your deliberative pattern might be simply to switch into the reactive "following" pattern at the conference, to engineer the opportunity.

Real-time deliberative patterns are more challenging than preparatory deliberation. Each of the examples below is either a fragment of real-time deliberate simulation, or a reactive pattern designed to enable such a fragment:

- *If-then*: Suppose that in a job interview, a candidate has answered the classic "What are your weaknesses?" question with "I sometimes work too hard." You might simulate potential follow-up questions using loose if-then reasoning: "If I call him on his clichéd avoidance tactic and ask for a real weakness, I might learn about his honesty and sense of humor or he might stick to his position. But if I pointedly ignore the answer and change the subject, he might sense he has lost some credibility and be on the defensive for the rest of the interview."

- *Focusing*: Saying "You've raised several points, and I hear a few different themes in what you say; let me address them in order." This deliberative meta-tactic uses a simple reactive tactic to simplify an anticipated fragment of deliberation. Tactics used: aggregation, separation, shaping, sequencing and selection. Underlying source domain for metaphor: manipulating a collection of many small objects.

- *Deliberative Dominance*: When you are in a situation where you have a decisive advantage in setting the agenda, such as being an interviewer rather than an interviewee, you can restrict the other person to predictable reaction patterns, while limiting your own reactive behaviors to a minimum. The deliberative element here involves pre-defining a landscape of ideas, a set of preferences about where you want the conversation to go and which subjects you want to avoid (a spatial and flow metaphor). In the live situation, the (primed) reactive pattern is to let the map in your head dominate the situational realities, the way a set of dams, channels and valves controls the flow of water in a river valley. The American conservative television personality, Bill O'Reilly, practices such adversarial interviewing to an extreme and literal degree, often demanding of his studio staff that they "cut his mic" when an interviewee resists less-assertive guidance.

- *Delayed reaction*: In the sense of the military phrase "wait till you see the whites of their eyes" (attributed to William Prescott, during the American War of independence), this is a particularly interesting phenomenon, since it merely modulates a basic reactive pattern by adding a single deliberative element: a time delay. A reactive pattern is usually enacted immediately after the stimulus pattern is noticed in the environment. The default is to push back when pushed. Delaying in order to time a reaction for maximum effectiveness turns a reactive pattern into a deliberative one. What separates it from an immediate reaction is that deliberate timing requires some off-stage logic, to compute the required waiting time.

- *Recognition-primed deliberative action:*[39] During enactments, you develop a theory of the situation, at which point reactive behaviors give way to deliberative behaviors. When the theory of the situation matches one that you remember from previous experiences, you may choose to steer your current enactment with reference to the older, remembered enactment. This memory may be unconscious (even entirely in muscle memory), and derive from a training exercise rather than a previous live situation.

Deliberative patterns are most often associated with calculative rationality. Recognition-primed behaviors illustrate how narrative rationality can arise from calculative rationality.

Experts, characteristically, can skip quickly to correct answers without exhaustively exploring a complex decision tree of formal contingency plans. Effective recognition-primed behavior emerges when an inefficient and time-consuming calculative-rational decision tree is reorganized as a more compact but cryptic narrative-rational one.

This happens as a novice gains experience and insights that allow him or her to gradually incorporate the cheap tricks (Chapter 4) from every new experience into canned prescriptions.

Experts often describe novice behavior as being "by the book." Their own repertoire of cheap tricks and hacks can be understood, within this metaphor, as scribbled margin notes that improve a calculative-rational tactical manual (a literal plot element in *Harry Potter and the Half-Blood Prince*).

In some cases, a repertoire of cheap tricks can be codified into a more formal and teachable domain doctrine, or set of design principles. In other cases, they remain too deeply buried in muscle memory or the subconscious to even bubble up to a margin-scribble level of awareness.

Recognition-primed decision-making has been extensively studied by Gary Klein.[39] Malcolm Gladwell's *Blink*[40] is an entertaining popular treatment of the ideas.

5.4.3 Opportunistic

Extended enactments can span multiple situations, separated by situations that are part of other enactments. The strict sequentiality of time forces us to weave many ongoing enactments together.

When an ongoing enactment is interrupted by a situation that is relevant to a different ongoing enactment, currently in suspension, you get opportunism.

For example, you pass by the grocery store on your way home from work, remember you are out of coffee, and stop to pick some up outside of your normal grocery shopping routine. By interrupting one enactment briefly to enact a fragment of another, you make an opportunistic gain. If you run out of coffee on Sunday, and normally go grocery shopping on Saturdays, you can now avoid both an expensive extra trip to the store *and* several days of interruption of your morning coffee routine.

The keys to opportunism are *unconscious* recognition of a situation from one enactment while being absorbed in another, and the ability to artistically organize your enactments according to interval logic (the skill we encountered in Chapter 2). The two enactments may be loosely related, such as meeting a colleague from one project in a meeting about a different project, hardly a pure coincidence. Or they may be nearly unrelated, such as finding in the seat pocket of an airplane, a copy of an obscure book that you've been waiting to read.

Opportunism involves the same dynamic as a cheap trick in a deep story: recognition of the potential for disproportionate rewards, which is why you get a weaker version of the same "Aha!" feeling.

In general, when trading off the costs and benefits of one ongoing enactment against those of another, there are no bargains to be found. Few synergies are truly powerful. A no-free-lunch condition normally holds: gains in one enactment will be offset by comparable losses in another. At the extremities of the tradeoff, where we entirely abandon one enactment in favor of another, we normally think in terms of opportunity costs.

Opportunistic tactical patterns violate no-free-lunch constraints by exploiting local conditions (such as the fact that there is a grocery store on your way home).

Since time is often the resource under contention, so long as there is sufficient slack in the locally prevailing interval logic, unintended consequences due to opportunism are unlikely. You would not stop to buy coffee if you were rushing to catch a plane, but a five-minute delay getting home is usually acceptable.

For this reason, opportunism can often be catalyzed by introducing some sort of inefficiency into all your enactments. Temporal inefficiency (schedules with slack) is the simplest kind, but any sort of inefficiency helps. Highly optimized behaviors are also highly blinkered behaviors.

Whatever the level of serendipity, opportunistic behavior is driven by a more or less unconscious act of recognition, based on a suspended mental model.

Unlike deliberative recognition-primed behavior, to which it is related, opportunism involves significant improvisation. In the coffee example, you might actually drive by a grocery store that is not the one where you usually shop. Your normal shopping pattern, based on assumptions about the layout of aisles and the location of the coffee section, will need modification. The modification will usually involve a reactive or exploratory (natural) behavior. For example, you might wander through the aisles of the unfamiliar store randomly, hoping to find the coffee quickly (exploratory) or you might systematically zig-zag your way through all the aisles (reactive).

This element of necessary improvisation makes opportunistic decision patterns the most demanding and creative kind. This is one reason William Gibson's portrayal of Wintermute, an artificial intelligence program in his classic cyberpunk novel *The Neuromancer*, is so effective. Wintermute reflectively explains its behavior to the human protagonist of the story: "I try to plan, in your sense of the word, but that isn't my basic mode, really. I improvise. It's my greatest talent. I prefer situations to plans, you see..."

5.4.4 Procedural

Procedural patterns are the simplest to recognize but the hardest to explain. A procedural pattern occurs when a decision-maker engages successfully in a complex and highly effective enactment without understanding the logic of his or her own behavior. This requires an artificial environment, containing externalized mental models created by others. You may, for instance, use a formula to compute the answer to a problem without understanding how it works. A bureaucrat may enforce compliance with a regulation, and a customer of the process may comply, without either of them understanding the logic of the paperwork involved.

This is perhaps the most subtle sort of decision-pattern, and can only be deployed when the prevailing mental model has been externalized into the environment through the processes of codification and embedding of enactments.

This implies the existence of something like external systems and processes (such as *In* and *Out* trays on your desk). These in turn self-assemble into organizations.

We will defer detailed discussion of procedural patterns to Chapter 6. For now, you can think of procedural patterns as natural behaviors being transformed into one of the other patterns (reactive, deliberative or opportunistic), without much thought required of the agent. The information driving the behavior has been consciously engineered into the environment in a way that is only partly comprehended by the person being influenced.

5.5   Enactment Styles

The major claim in this book is that tempo can be used to modulate enactments that move along too quickly to manage one micro decision at a time. A consistent approach to such modulation is an *enactment style*.

An enactment style combines a domain-specific aesthetic with a doctrine (Chapter 3), in a given situation. Your enactment style will be strongly influenced by your doctrine, but will not be completely defined by it. It will depend on the situation and domain. The primary signature of an enactment style is its characteristic tempo.

For example, if you are a seasoned tennis player, prefer your backhand to your forehand, and your doctrine is *Ready-Fire-Aim*, then your primary enactment style in the tennis domain is likely to be aggressive, and heavily reliant on your backhand. But you might adopt a different style in a specific match, or in special situations such as coaching. In domains that reward patience more than aggression, you might adopt an unfamiliar style, even if it feels awkward.

We need to define a unit of narrative time (Section 4.5) before we can proceed. An enactment *epoch* is a narrative-time period of indefinite length, characterized by a consistent enactment style. So any string of basic decision patterns and isolated tactics having a distinctive tempo is an enactment epoch. I'll simply refer to this as *epoch* from here on, but keep in mind that narrative time, unlike calculative-rational "container" time, does not move forward unless an enactment is actually progressing.

Try reviewing your calendar over the last year to see if you can detect the tempo-changes that mark the beginnings and ends of epochs. If you've been trying tempo-doodling (Chapter 1), you should have already developed a sensitivity to epoch changes.

The Double Freytag pattern we encountered in Chapter 4, by this definition, has seven distinct epochs. Two of the six tempo transitions, the cheap trick and separation event, are marked by special events.

In our analogy to language, where basic decision patterns are like sentences, epochs are like longer passages, written with a consistent style.

At the level of individual universal tactics and basic decision patterns, domain effects significantly constrain enactments. There is little room for the expression of an enactment style, since every tactic or pattern you employ must be chosen primarily to cause a desired effect.

There is, however, room for variation, both in your choice of patterns and tactics, and in your enactment of those choices. To move an object, you may choose to push it along the floor, or pick it up and place it where it needs to go. If you decide to push, you can choose to push gently or forcefully.

The *consistency* you exhibit in your variations reveals your enactment style. Style is always a cumulative phenomenon revealed through repeated reinforcement. The specific narrative clock at work determines the minimum clock-time length of an epoch required to clearly reveal a style. A boxer may reveal his style in a few seconds. A new CEO may require a few quarters. In narrative-time terms, however, the two epochs may be the same length. For instance, six punches for the boxer and six meetings for the CEO.

Mathematically, a style is a statistically significant pattern within an enactment, and the length of the epoch required to establish significance is a function of the complexity of the pattern that defines the style, and the rate at which raw information is being generated in an enactment.

In writing, paragraphs and passages represent the scale at which you can effectively reveal (or attempt to disguise) your chosen style. Principles such as using short sentences, avoiding the passive voice or avoiding adjectives are what create specific stylistic effects. When a writer pays more attention to these style effects than to the content of what he or she is saying, we remark that the writing is highly *stylized*. When there is a mismatch between the information content and the refinement of the style, the writing appears to have an overwrought style-over-substance character.

A single short sentence does not have much of an effect. But consistently choosing short sentences, page after page, creates a very definite rhythm in the text.

A single sentence in the passive or active voice has little impact by itself. But consistent use of the passive voice lends the text a languid feel, while consistent use of the active voice lends it an energetic feel. By modulating the mix of active and passive voice, a writer can modulate the overall energy of a text.

And finally, a single sentence that lacks adjectives does not appear unusual. But an entire passage containing no adjectives would create a very definite emotional effect (you might call it Spartan prose without quite realizing why it feels that way).

These choices in writing *cumulatively* affect rhythms, energy and emotions: they are about tempo.

Decision epochs with a given enactment style work exactly the same way.

Each of our four basic types of decision patterns (reactive, deliberative, opportunistic and procedural) represents a pure style. A bureaucratic enactment style, for example, is marked by near-exclusive use of procedural patterns, while an improvisational style is marked by heavy use of opportunistic patterns.

At the more fine-grained level of universal tactics, consistent choices usually come across as subtle flavors that help distinguish among enactment styles. We commonly use adjectives inspired by the source domains of particular conceptual metaphors, such as *mechanical, fluid* or *heavy-handed* to characterize such flavors.

Let's look at an example.

5.5.1 Butterfly-Bee

Muhammad Ali's famous *float like a butterfly, sting like a bee* approach to boxing is an enactment style that mixes all four kinds of patterns elegantly. There is a deliberative element (the decision to move around rapidly) combined with reactive-opportunistic patterns (moving in response to, and faster than, the opponent, while looking for opportunities to "sting"), all overlaid on a vocabulary of procedural patterns (the basic by-the-book punch combinations that boxers inherit from whatever stylistic tradition drives their training).

Butterfly-Bee is a style that relies primarily on pace-setting and pace-disruption (Chapter 2) via *acceleration*. It is perhaps the simplest illustration of the philosophy of decision-making known as the OODA loop (observe, orient, decide, act) developed by John Boyd.[7]

Despite the stuffy and bureaucratic-sounding acronym (and the forbidding process diagrams often used to describe it), OODA is actually the core of a deeply creative and philosophically elegant enactment style that is based on many of the same themes inspiring this book: tempo, entropy and creative-destruction. Boyd's overall philosophy could be called a meta-style, since it shows up in so many domains. Butterfly-Bee is the manifestation in boxing.

The central idea in OODA is a generalization of Butterfly-Bee: to simply operate at a higher tempo than your opponent. This is a subtle point. A higher tempo is *not* the same as higher speed, in the sense of a race car overtaking another. To think in terms of tempo means to think in terms of (narrative time) frequencies rather than speed.

The primary effect of operating at a faster tempo is that you can maneuver inside the decision cycle of your opponent, disrupting his or her enactment by introducing entropy into it faster than it can be removed. In terms of the entropy-based concepts in Chapter 4, you are forcing your opponent to stall in the high-anxiety exploration epoch, or prematurely commit to the wrong cheap trick out of sheer desperation.

This style is obviously effective in adversarial settings, but it is equally effective in cooperative settings, a point that many who are inspired by Boyd miss. The alert waiters at upscale restaurants, fulfilling your needs before you even recognize them, are a good example. Exceptional customer service, not just war plans, can arise from OODA thinking.

5.5.2   Hierarchical Dynamics

In this non-adversarial form, "faster tempo" is the principle underlying the effectiveness of well-aligned hierarchical organizations. Higher-level enactments in a hierarchy achieve control through modulation of faster lower-level enactments.[‡] This is the idea underlying all hierarchical governance. Hierarchies are naturally organized to operate at lower tempos at higher levels. This insight is usually mangled and misunderstood in a deeply flawed way: that higher-level enactments are about longer time horizons. Horizons are actually determined by predictable epoch shifts (changes in tempo), rather than tempo itself.

The distinction between adversarial and non-adversarial situations is also what enables guerilla warfare and intra-organizational rebellions. Consider a nominal authority, such as the management of a corporation or a military occupation force, that is interested in hiding conflict and dissent behind a facade of governance by consent. Such an authority, to appear credible, must necessarily act through the hierarchical models that make sense if consent and consensus have actually been achieved.

This provides lower levels in a hierarchy with a natural channel for disruption: creating misleading patterns of information overload at higher levels, or using the *fait accompli* and other momentum-based tactics (Chapter 3) to force commitments.

We will not have time to examine hierarchies in detail, but I write about this topic frequently on my blog, ribbonfarm.

5.5.3   Information Flows and The Effectiveness of Styles

Styles like Butterfly-Bee, derived from the Boyd OODA meta-style, are most effective when the primary direction of *privileged* information flow in a domain is bottom-up. Combat is an obvious example of such a domain. When the action starts, the fog of war blinds generals before it descends on privates.[§]

Generalizing this principle, the effectiveness of a given style, with its characteristic principles of pace-setting, pace-disruption and momen-

---

[‡]Electrical engineers will recognize a very attractive metaphor here, based on interference and amplitude modulation effects in signal processing.

[§]This is literally true in some simple domains. I once created a rather fun computer simulation to demonstrate this principle; email me at vgr@ribbonfarm.com if you'd like the code.

tum management, depends on the pattern of privileged information flows in a domain.

In a hierarchy, passive aggression works best when the primary direction of privileged information flow is outward from a privileged, mid-tempo middle. The tempo at which the Federal Reserve manages interest rates is an example. To be effective, this tempo must be *slower* than that of the market being managed, and *faster* than the legislative authority that oversees the Federal Reserve's actions.

Finally, autocratic regimes work best when the primary direction of privileged information flow is top-down. Apple is an example of a company that is defined by such flows, since the company has been based on the preternatural design and market insights of Steve Jobs.

The British television comedy series, *Yes Minister* and *Yes, Prime Minister*, showcased many entertaining examples of both middle-out and top-down styles.

5.6   Skill: Thriving in the Strategy-Tactics Fog

The words *strategy* and *tactic* are ambiguous and foggy to the point that many impatient decision-makers mistakenly dismiss them as vacuous. Much of this confusion is due to the poverty of the language of calculative rationality, and the fact that people are tempted to seek prescriptive definitions where none exist.

If we are more modest in our aspirations, the vocabulary of narrative rationality allows us to craft a pair of simple but cryptic and non-prescriptive definitions.

- *A strategy is a cheap trick.*

- *A tactic is a metaphoric mapping among primitive action concepts in two or more domains.*

These definitions have taken us several chapters' worth of work to get to, but they do precisely nothing to make the underlying creative-thinking challenges easier. So we must ask: why bother with definitions at all, if they are non-prescriptive?

The advantage to operating by them is like the advantage to be found in knowing that perpetual motion machines are impossible. The knowledge allows you to avoid certain patterns of failure. But knowing how not to fail is not the same as knowing how to succeed.

All these definitions do is constantly remind you that the creative elements in decision-making – novel perspectives and novel conceptual-metaphor mappings – are locked in a yin-yang embrace that cannot be engineered away by any calculative-rational formula.

In writing for example, great stories stretch the limits of a language, and an evolving language constantly challenges its storytellers. The result is a living literary tradition. To limit yourself to dictionary meanings, formulaic plot devices and bureaucratic notions of grammar is to imprison a language within a few tired genres and style guides.

Decision-making is ultimately about constantly looking for the next cheap trick in ongoing enactments, scripted in constantly shifting tactical vocabularies. You will occasionally get a temporary and local advantage, but ultimately nature will defeat you.

You can, however, consistently outmaneuver those who *don't* recognize that there are no universal formulas, just as you can consistently outplay someone who always plays "rock" in a game of rock, paper, scissors.

The equivalent to always playing "rock" in general decision-making is always reducing the fluid, creative-destructive strategy-tactics distinction to a specific rigid and prescriptive one. Each such reductive distinction creates a particular pattern of confusion.

Each of these patterns of confusion (I am tempted to call them "patterns of strategery" in honor of former President Bush) is characterized by *repeated and consistent* use of certain impoverished distinctions, and occupation of preferred positions based on those distinctions. Here is a list of the most common ones, along with examples of the characteristic language that signals the presence of each pattern:

- *Levels and Horizons*: "Strategy is about the big picture/long-term; tactics are about the details/short-term."

- *What vs. How*: "Strategy is about *what* you want to do; tactics are about *how* you do it."

- *War vs. Battle*: "Strategy is about winning the war; tactics are about winning individual battles."

- *Why vs. What*: "Tactics is about *what* to do; strategy is about *why* you should do it."

- *Amateur vs. Professional*: "Amateurs worry about strategy and tactics while professionals worry about operations."

- *Process vs. Goal*: "That's a process question, we'll work that out later; let's establish our intent first."

- *Abstract vs. Specific*: "Let's consider this at a more abstract level" or "The devil is in the details."

- *Global vs. Local*: "We have to think globally, not locally," (variant: technically trained people like to substitute "optimize" for "think").

- *Specialist vs. Generalist*: "Our problem is wooly headed generalists," or "our problem is narrow-minded specialists."

- *Conceptual vs. Technical*: "Forget those technical details; what's the *real* conceptual question here?"

- *Systems Thinking vs. Silo Thinking*: "We need more systems thinkers around here; everybody here is trapped in their own little silo."

Note that each of these distinctions can actually be useful in specific circumstances. That's what makes them tempting. The symptom that you are dealing with a pattern of confusion is that the *same* distinction keeps popping up in *every* situation, with individuals consistently occupying the *same* position with respect to that distinction. When such patterns take hold, you generally get escalating battles of "I am more big-picture/detail-oriented than thou" among those who compete for the same position, and mutual contempt among those who have opposing preferences.

This list is obviously not exhaustive. The era of canny or deluded inventors pitching designs for perpetual motion machines to gullible investors may be over, but the cottage industry of strategy gurus selling perpetual creativity machines to gullible buyers continues to thrive. So there are always new patterns of delusion to detect and exploit.

There are two ways to exploit each of these patterns.

First, it is always possible to directly challenge and arrest the progress of a useless distinction before it gathers too much momentum, by using a situationally relevant counterexample. Typically one that either complicates the simplistic distinction, or exaggerates the distinction to the point of absurdity. For example, in the world of business you can challenge the "levels and horizons" distinction with "but we have a strategically critical presentation coming up tomorrow!"

This sort of open challenge approach, however, generally leads to tedious, thankless and ultimately doomed projects of language reform. The lure of formulaic creativity is generally irresistible, and it is rarely possible to win over devotees of a particular formula.

Sadly, it is generally more productive to reinforce individuals' preferred distinctions and positions by appealing to their vanity, and using their immobility to make your own search for creative inspiration a little easier.

Definitions are unfalsifiable constructs. The success of a new definition ultimately rests on the accuracy with which it manages to label an intuitive and useful, but unarticulated concept. This success must be measured in terms of the fertility of the thoughts and questions it inspires, during whatever brief reign it enjoys.

So I will leave you with a few interesting questions which have proved productive for me. I hope these prove productive for you as well.

1. Are all tactics derived from universal tactics, or are there tactics that represent direct mappings among non-universal domains?

2. Is there any such thing as a pure formula confined to a single domain, that cannot metaphorically translate to another domain?

3. What distinguishes true cheap tricks within deep stories from Aha! moments in simpler enactments?

4. Does the leverage provided by a given cheap trick diminish through reuse in subsequent enactments?

5. Does this leverage diminish rapidly enough that there is no value in generals being prepared to fight the last war?

6. Do forgotten historical cheap tricks regain their vigor and leverage after a period of abandonment?

7. Is it possible to systematically improve your sense of history, by approaching the subject in specific ways?

8. Is a cheap trick that has been recognized by an adversary automatically rendered useless?

9. Is a strategy shared no longer a strategy?

10. How does the efficient market hypothesis apply to the market of cheap tricks diffusing within and across domains?

11. How quickly do cheap tricks diffuse within and across domains?

12. Can we compute the time bought by a cheap trick *a priori*?

Chapter 6

The Clockless Clock

To see a world in a grain of sand,
And a heaven in a wild flower,
Hold infinity in the palm of your hand,
And eternity in an hour.

From *Auguries of Innocence* by William Blake

We're nearly at the end of our uphill hike. Take a moment to stop and look back. We've charted a rather barbaric course through some very civilized territory. On occasion, we've dismissed centuries of scholarship – around formal models of planning, for instance – with a casual sentence. On other occasions, we've paid unusual amounts of attention to ideas, such as Archilochus' thoughts on fox and hedgehog archetypes, that might seem at best peripheral and at worst whimsical and irrelevant. We have ignored certain subjects, such as probability theory, that are central to calculative-rational decision-making, while devoting what might seem like inordinate attention to distinctions, such as the one between *strategy* and *tactic*, that many dismiss as a mere semantic worry.

That is the cost of adopting a new perspective. Proportions change, and ideas appear in different relations to each other. What was once at the forefront is hidden and obscured, while peripheral elements occupy center stage. Major ideological divides are spanned with rough-and-ready conflations, while fine, seemingly fussy distinctions in the older perspective widen into yawning chasms.

The sole justification for such intellectual vandalism is our hope that a new perspective will help us understand the world better and let us do things the older outlook could not.

In this chapter, we will attempt to do both. In the process, we will also get to the edge of what we currently know. Our goal will be to apply the ideas we have encountered so far to the world beyond the individual: groups, organizations and the fabric of social reality that we create for ourselves.

But this goal will be secondary and nominal. What I really hope to do is increase your sense of comfort and familiarity with the core ideas, and to arm you with the right mix of confidence and skepticism that will allow you to tastefully adopt this perspective as your own, should you wish to do so. In other words, I hope to help you develop a sensibility, not teach you an idea. I call this sensibility the *clockless clock* sensibility.

The true test of the power of a new perspective is its ability to make sense of things that were not explicitly considered before its adoption. So far in this book, though I have framed things in unusual ways, and on occasion offered provocative views, I have mostly stayed on firm ground. I have mainly talked about things that I actively considered during the formulation of the core ideas, things that I set out to explain.

In this chapter, that will change. We will attempt to test these ideas far afield from where they were grown.

We must make our way from the firm ground of broad consensus, through the marshes of dissent, to the treacherous shifting sands of conjecture and unanswered questions. We will only be able to sense what lies beyond as a vague sense of uncertainty, since the shoreline of the unknown-unknown void out there is by definition invisible.

6.1   Externalized Mental Models

The four central ideas we have encountered – timing, mental models, narratives and metaphor-based universal tactics – allow us to look at the world from a certain perspective: everything is a mental model.

This is trivially true of natural realities: we can only comprehend them through mental models. It is also true of artificial realities, but in a less trivial way. The material things we create individually and collectively are *externalized* mental models: parts of reality arranged to conform to the structure of an internal mental model of the sort we discussed in Chapter 3.

Kant once said, *we see not what is, but who we are.** There is more truth to this sentiment than you might think.

6.1.1   Codification and Embedding

A to-do list is an externalized list of intentions; we make to-do lists on paper because that's easier than remembering them in our heads. As we saw earlier, intentions are among the three primitive building blocks of mental models (the other two being beliefs and desires), and a to-do list is one of the simplest and most obvious kinds of externalizations.

We externalize mental models because they are burdensome. And we are not alone in the natural world in being lazy thinkers. Daniel Dennett, in *Consciousness Explained*[41] describes a strange creature called the sea squirt that has a primitive brain in its juvenile state. It uses this brain to swim around looking for a rock or a piece of coral to cling to. When it finds one, it "doesn't need its brain anymore, so it eats it!"

Here is a more complex example: how we organize our desks reveals a great deal about our mental models of work. Our desks encode our beliefs, desires and intentions. They reveal the possible worlds we are considering and even elements of the deep stories we are living through. An austerely clear, Bauhaus High-Modernist desk represents a doctrine of clinical efficiency, while a desk with lots of desk toys and cartoons represents a more relaxed doctrine.

To externalize a mental model is to arrange our physical environment to serve our intentions more efficiently. In cognitive psychology, this idea is called *distributed cognition*. Those familiar with distributed cognition will recognize many of the concepts we will consider, though they may appear in a different guise and with a different pattern of emphasis.

A to-do list involves language. Here is an example that does not.

Suppose I believe being calm and relaxed at work is good for me, and that having plants makes me more relaxed. So I put a potted plant on my desk. If it works, I no longer have to think about it: I can devote fewer mental resources to maintaining those two beliefs.

Externalization happens through two steps. The first step is *codification*: expressing part of the meaning of a mental model in a form that can be manifested in material terms. In the case of the to-do list, a set

---

*I am not able to track down the source of this quote

of (possibly nonverbal) intentions is codified using a human language, a "put it in words" step that precedes the actual writing. In the second case, beliefs about plants and emotions are codified into a vision of a future state of your desk.

The second stage is an embedding: creating physical consequences that endure. A paper to-do list might last a week before being recycled. A potted plant might last years. A sentence in a human language, even if unspoken, is a codified thought. If spoken, it might embed itself in the memory of a listener. It is still an act of externalization because the other person's mind is external to yours. This is why the punch sequences learned by a boxer (an example we encountered in Chapter 5) constitute procedural knowledge: the knowledge of a coach or teacher being externalized into a student's mind.

These are trivial examples. The differences among mental models, codification and embedding gain more significance when we consider larger scales.

For example, an architect's private vision of a future city might be codified and embedded as a set of engineering plans for one audience, a speech for another, and the actual completed city for a third. The plans, speech and the actual city are all physical manifestations, or embeddings, of elements of the mental model, that emerge through different codification processes.

Note that a codification may remain within a mental model as a possible world, without being embedded at all. You may think through an idea for a speech that you never give, entirely in your head. But in general, a codification is best understood as a process that leads on to an embedding.

Recall our analogy, in Chapter 3, between mental models and weather systems. Within that analogy, codification and embedding are rather like precipitation: crystallization eventually leading to the deposition of moisture on the earth's surface in some form. Codification, like cloud-formation, may or may not actually lead to precipitation.

6.1.2   Internalization and Communication

When we study nature, we make up mental models based on hypotheses about cause and effect. We attempt to infer structure and law in what we see.

Modern humans who are comfortable with Darwinian thinking do not read agency and design into what they see in nature, or attempt to make sense of a hypothetical agent. Or as philosophers say, we do not appeal to teleological explanations (viewing the behavior of systems as being driven by agents seeking specific desirable outcomes).

When we study artificial realities, however, we *do* seek to understand behavior in terms of agency and design. We correctly assume that artificial realities are easier to understand in terms of the mental models of those who created them. So, to a lesser or greater extent, we *internalize* the original mental models of those who externalized them. Consider this to-do list:

- Call cat-sitter to confirm

- Get oil change

- Set Out-of-Office message

- Check-in online

- Pack

- Remember to get gift for niece

It is obvious that this is a list made by somebody about to take a trip. The one inconsistency should also jump out at you: getting an oil change *and* checking-in. Is the person driving or flying? Perhaps there are two people, one of whom is flying? Or perhaps the person is driving part of the way, flying the rest.

Thanks to a simple to-do list, you've *internalized* a fairly complex mental model.

Let's extend the potted plant example from the previous section. Imagine that a successor inherits my desk and potted plant. He has no opinions on plants, and keeps the plant as a default decision, watering it occasionally. Despite himself, he finds that he feels calmer, and likes the feeling. If he is introspective, he might infer and adopt the same beliefs: *being calm and relaxed at work is good for me*, and *having plants makes me more relaxed*.

If he is *not* introspective, and does not update his mental models about desks, calmness and plants, then he has ended up in a state like Dennett's sea squirt: enjoying an advantage, but with no idea why.

Consider two other possible reactions.

First, my successor might notice the calming effect and *not* like it. He might believe, "I shouldn't calm down too much, I need to be alert and have a sense of urgency in this job." So he might accept one belief (plants are calming) and the opposite of the other (being calm and relaxed at work is *not* good for me). In this case, a different doctrine triggered a different internalization.

And finally, a skilled interior designer or psychologist looking at my desk might internalize a much richer mental model than the one I externalized, perhaps inferring the effects of repressed, traumatic childhood experiences, that are invisible to me.

An internalized mental model, therefore, will share some information with the mental model that was originally externalized, but may differ in significant ways. It may be either richer or more impoverished in terms of the sheer quantity of information it encompasses.

This process of mental models triggering related mental models in other minds is what we understand as communication in its most general sense.

The potted plant is effectively a (non-linguistic) message transmitted from one (possibly long-dead) person to another. This is the reason anthropologists studying long-dead cultures view *any* archeological finds as "texts," whether they are talking about garbage, shards of pottery, or inscriptions in a strange dead language.

Spoken language is actually the hardest sort of communication to comprehend within this model of externalizations triggering internalizations. The material medium of the embedding is the air between speaker and listener. Both externalization and internalization must occur in this most temporary of all media. This can work because both speaker and listener share a common frame of reference: if I say "apple" to you, I am taking advantage of two older, slower communication acts in our pasts, when we each, as children, first had physical apples pointed out to us.

When you attempt to learn a new language, you must again start with such basics as pointing at apples.

6.1.3 Seeing and Creating

Internalization and externalization are merely nuanced ways of under-standing the familiar processes of seeing and creating, with respect to artificial environments. The two are intimately linked. How we create depends on how we see, and how we see depends on how we create, as suggested by the familiar witticism, "when you have a hammer in your hand, everything looks like a nail."

Human creations are an outcome of the chemistry between what we actually see, and the simpler conceptual categories in our heads, such as circles, straight lines and right angles. Children start by drawing stick figures based on simple platonic shapes. Realistic drawing is a learned skill for most of us; we must consciously practice suspending our natural conceptual filters.

There exist brain-damaged individuals who can *only* view the world through platonic shapes, and are unable to recognize natural objects such as faces and flowers. There also exist savants who can draw with perfect realism with no training. Oliver Sacks has explored both ex-tremes in *The Man Who Mistook His Wife for a Hat*.[42]

To operate in a complex world, we need both the capacity for see-ing through simplified categories and the capacity for seeing reality as it actually is. In humans, a mechanical, photographic way of seeing and creating is actually a liability. We may marvel at savants with photoreal-istic drawing abilities, but this wonderful capacity is part of a condition that makes them incapable of functioning as autonomous adults. Those who can only see the world as a set of platonic shapes are equally inca-pable of functioning as autonomously as the rest of us.

You could therefore describe normal human thought as a balancing act between sensing realistically, and seeing through simplified platonic abstractions.

When we create, our work usually reveals a bias towards one side or the other. The more we desire control and comprehension, the greater the extent to which the realities we see are simplified by the platonic categories of our mind, before emerging as creations at the other end. The more comfortable we are with chaos, the more we act as trusting, unconscious (and possibly uncomprehending) agents of raw perception.

These two extremes lead to a common pair of opposed archetypes of action in mythologies around the world: design and dance.

6.1.4   Design and Dance

Design is the detached, objective mode of action, represented in mythologies by gods of divine imperial order and wisdom, such as Zeus or Indra. They represent the dominance of platonic thought over perception; they are causal forces that transform without being themselves transformed. The pyramids of ancient Egypt are beautiful examples of our primal platonic design urges.

Dance is the archetype for a participatory mode of action, represented by the gods of ritual ecstasy and chaotic turbulence, such as Dionysus or Shiva. These represent a personification of natural forces of change, channeled through an unconscious framing aesthetic that manifests itself as ecstatic dance. The dancers are themselves microcosms of the transformations they bring about.

Design and dance are enshrined in our collective mythological imagination as the two primary drivers of action. For philosophers and poets, they lead to probing of such metaphysical profundities as the "being vs. becoming" duality. What we need to understand, though, is how more prosaic forms of seeing and doing are affected by these ideas.

Whatever balance we strike between design and dance in our actions, the externalized mental models that result are generally more *legible* to other humans than natural realities. It is immediately clear that there is an artificial, human-caused order to things; that there is human meaning embedded in the environment. Unlike William Blake, who wondered "what immortal hand or eye could frame" the "fearful symmetry" of the tiger, when you and I look at a desk, a park or a building, we can comprehend the human creative force behind it.

Even a document in an unfamiliar language exhibits a sort of surface legibility in a way that the image of a tree does not (the meaning of the tree, to the extent that it has one, is down at the DNA level). This association between legibility and human-like agency is so strong, and so generally valid, that it can lead us into interesting mistakes, such as the 19th century belief in "canals" on Mars (which later turned out to be optical illusions).

6.1.5   Legibility and Meaning

We are using the word *legible* in a slightly deeper way than is usual here. Typically we mean something like *readable*. In this usual sense, *illegible* corresponds to something like a smudged piece of written text. The original text is the result of coherent human intent (which we recognize even if we do not understand the language) while the smudging is an accident from the human perspective. The only order in the smudge is due to the laws of physics operating on ink and paper.

Our use of the term is similar but broader: a piece of physical reality is legible if it is obviously the product of coherent human agency, a deliberate externalization of a mental model. When human and natural sources of order are harder to tease apart, you get greater illegibility. This deeper notion of legibility was developed by James Scott in his fascinating study of the mental models involved in failed, centrally planned public policy projects.[43] Using our vocabulary, Scott's failure model can be described as a case of overweening impoverished design crushing an ongoing organic dance.

Illegibility can be a result of intentional obfuscation, as in the case of camouflage. In the Sherlock Holmes short story, *The Adventure of the Dancing Men*, a gang of criminals uses what look like childish stick-figure doodles as a secret code. But Holmes recognizes language lurking underneath the camouflage.

In other cases, illegibility can arise from sheer complexity. The structure of an old city is the result of generations of residents gradually modifying their environment. It is the *collective* externalization of the partially shared mental models of millions of residents over centuries. It will therefore be *necessarily* illegible to a single urban planner, who will fail to completely internalize the collective emergent externalization, and view what he cannot internalize as chaos; the effect of natural "smudging."

In Scott's model, central planners begin by poorly internalizing a rich, emergent collectively externalized mental model (such as the structure of an old city). They proceed inevitably to failure when they attempt to replace it with the externalization of their own impoverished mental models, constructed largely out of primitive platonic ideals, a case of design destroying dance. Used with adversarial intentions, Boyd's OODA can be understood as a deliberate use of illegibility to cause failure. As Scott notes, rebellions often start in illegible parts of cities: slums.

Collective, emergent mental models of the sort studied by Scott are at an intermediate level of legibility. They are not quite as illegible as the most obscure workings of nature, but they are not as legible as the desk of a single individual. The products of deliberate design, strongly driven by the categories of a single mind, such as a chair, are highly legible. The products of emergent, *laissez-faire* design, such as the economy at large, can be nearly as illegible as natural realities. Deciphering the meaning behind artificial realities can be as simple as reading the label on a drawer, or as hard as the most difficult fundamental problems in physics.[†]

But let's return to individual human scales and examine how legibility can lead to meaning.

Suppose I am working on several projects at once. I have on my desk a messy, mixed pile of papers related to all my projects. I believe it would help if I were to organize all the paper materials in my office by project.

I could do this in two basic ways.

First, I could get a set of binders, label them neatly, and file the papers by project in alphabetical order.

Alternately, I could just stack my papers in big, unwieldy piles, one pile per project, on the floor, with the papers roughly ordered according to the internal logic of each project.

In both cases, I would have externalized my mental model, and simplified the organization, by approximately the same amount. Labeling the project folders and imposing an alphabetical order is not obviously better or worse for me than unlabeled piles in a more idiosyncratic order.

In the first case, searching might be slightly harder; I'd have to read labels and search alphabetically, and alphabetical ordering may not be the most natural one for my papers.

In the other case, I'd remember something like "that pile near the window has to do with Project A, and the ones near the bottom are the ones I inherited from my predecessor." Searching might be faster, but I'd have to maintain an intuitive spatial map of the organizing scheme in my head.

But there *is* a big difference between the two ways when other people are involved. While both approaches to organization are legible to you and me, the latter is *meaningful* only to me. You will have to work

---

[†] When asked why humans could discover atomic power, but not the means to control it, Einstein reportedly responded "because politics is more difficult than physics."

hard to discover the meaning. I am the local, you are the outsider.

If you are my boss, and think like one of the dictatorial central planners described by James Scott, you might demand that I get organized in a company-standard way. I might even comply under duress. But chances are the imposition of a Procrustean-bed outsider mental model (whose main purpose is to make the situation legible and meaningful from an outsider perspective) might simplify my workspace to the point where I am no longer productive. You might ask me to remove my potted plant, for instance, because it is against a centralized decor policy.

This brings us to a familiar topic: organization.

6.1.6 Meaning and Organization

David Allen observed in *Making it All Work*[44] that to be organized simply means "where something is, is related to what it means to you." Sherlock Holmes, for instance, stored his tobacco in a Persian slipper. So long as *Persian slipper* means *tobacco container* to you, you can consider yourself organized if you store tobacco in one, even though your behavior might seem idiosyncratic to others.

This is a subtle point. Organization is about the meaning that matters to the person whose mental model is being externalized. Many people fail when they attempt to get organized because they make the mistake of striving for legibility and meaningfulness to an external eye, by imposing conventional or received social meanings onto personal realities:

- They might arrange their books by the Dewey Decimal system, believing it to be the "scientific" way, instead of a more useful logic for small personal collections (such as read/unread).

- They may arrange their kitchens the way grandma did, and cook the way she did, but fail to manage realities that grandma did not have to deal with, such as microwave ovens and frozen meals.

- They may strive for external legibility consciously: "I want my boss to see how organized I am." This means they focus on the peripheral (filing, labeling, stacking things geometrically). These aspects of organization are secondary and subservient to meaning. They may even be counter-productive when they don't align with the meanings in the significant mental models.

So apparent creative chaos is fine, so long as it is sufficiently legible and meaningful to *you*. The externalizations of your mental models only have to be legible and meaningful to others to the extent that you must share meaning with them.

We usually get organized in order to be more productive. To get at the relationship between organization and productivity, we need another idea.

6.1.7   The Tempo of an Externalization

Consider two artists, each operating by a mental model illegible to others, externalized into their respective studios. The two studios are approximately equally legible and meaningful to outsiders. Each represents a distinct style of creative chaos.

Yet when you and I walk in on our tour of the artists' studios, we both agree: there is a certain vigorous energy and a sense of passion about the first studio, while the second feels deeply depressing and stalled.

We are back at tempo: rhythms, emotions, energy.

Even when an externalized mental model is *not* particularly legible or meaningful to you as an outside observer, you can still detect the tempo that animates it. The simplest illustration of this is handwriting in an unfamiliar script;[‡] you can sense the rhythms, emotions and energy even if the script is an unfamiliar one.

The tempo of an externalized mental model is the reason why you can land in an unfamiliar and strange old city, and immediately get a feel for the place. Tempo cannot be easily hidden, which is why disguising it is usually the hardest aspect of deception. Adolf Eichmann, the Nazi war criminal, was finally captured in Argentina because an alert agent noticed his characteristically European saunter.

The first artist in our example is in a creative groove; his internal and external mental models have a lot of momentum and energizing emotion associated with them. The second artist is probably suffering from a creative block or depression.

We sense energy and emotion as elements of a gestalt. The overall

---

[‡]The signal processing methods applied to signals from outer space by projects such as SETI rely primarily on mathematical models of rhythms and energy to pick out candidate alien communications from the background noise. Emotion unfortunately is harder to work with, when dealing with non-human minds.

sense of vigor and energy in the first artist's studio might be triggered by several in-progress paintings, still-wet palettes, and framed paintings labeled for shipping. The second studio might be dusty, with dirty brushes that are crusted with dry paint (motivated, working artists are never sloppy about maintaining their equipment). Such obvious cues, as well as subliminal cues, contribute to the gestalt that we understand as tempo.

The tempo of an externalization allows us to develop situation awareness and manage momentum even in illegible and low-meaning situations, even as we begin the process of internalizing a mental model based on what we see. It isn't hard to walk into someone else's depressing, cluttered kitchen, full of dirty dishes, and take control, simply by starting to clear up. We can begin to act decisively even before situations become legible and meaningful to us.

The idea of externalization is the reason we haven't been too fussy about distinguishing between subjective and environmental tempos: in the human sphere, they are practically the same thing. Recall the four elements of timing we talked about in Chapter 2: merging, going with the flow, pace-setting and disruption. We can now read a new meaning into those dynamics. They represent the interaction of internal mental models with externalized ones. Mental models react with each other *even* when they are mutually illegible, because the tempos interact.

But there is one important difference between internal and externalized mental models. The primitive elements of internal mental models are beliefs, desires and intentions. Externalized mental models do not have such fine-grained primitive elements, since they are constructed out of physical materials. You cannot map a specific physical element, such as a coffee cup on my desk, to a single belief, desire, intention, possible world narrative, or deep story.

Instead, there are two coarser building blocks: *fields* and *flows*.

6.2   Fields and Flows

The externalization of mental models is a natural process, and causes a rich and varied phenomenology that is usually called *social reality*.[45] I will only consider the subset of social reality that can be understood in terms of two primitive elements, *fields* and *flows*.

Fields and flows are best suited to understanding parts of social reality that have a literally spatial aspect to them, such as buildings, cities, parties and armies. They are less useful when it comes to non-spatial social realities such as money and taxes. Field-flow thinking can be applied in these cases when strong conceptual metaphors are available,[37] but it takes more work. We will mostly restrict ourselves to literally spatial examples in this chapter.

Fields are arrangements of the physical environment. *Flows* are the behaviors that result when humans interact with a field. Flows are the observable parts of others' enactments.

Fields can be very simple: a coffee mug is an externalization of a set of beliefs, desires and intentions concerning hot fluids:

- Liquids spill less easily from tall containers.

- Hot liquids make containers hot.

- Humans do not like getting burned.

- Humans like drinking hot beverages.

- Handles make it easier to hold hot containers.

- A container that is much larger or much smaller than a fist is hard to handle.

- Humans generally drink fluids between 8 to 20 oz of any fluid at a given time.

- Containers with lips are easier to sip from.

Fields can embed a simple possible world, as in the case of the arrangement of paintings in a gallery, which encourages you to view them in a certain order: a flow encouraged by default. Some fields can embed an entire universe of deep stories, such as a college campus. Some fields contain a lot of momentum (in the sense of mental models, not physics), such as an assembly line. Others are relatively low-momentum blank canvases, such as an empty room.

Fields can be calculative-rational, such as a bureaucratic process, or narrative rational, composed of what Daniel Pink calls "emotionally intelligent signage," designed to accommodate the emotions associated with an enactment, rather than just the unadorned logical actions of a procedure.

Flows can be very simple: the main flow associated with a coffee mug is *lift, tilt, sip, put down.*

Flows can exhibit subtle dynamics. A mirror placed opposite a bank of elevator doors can turn impatient and restless pacing and repeated punching of buttons into a moment of calm self-absorption.

Flows can be orderly and highly controlled, such as a queue, or unruly and chaotic, such as a rioting mob egged on by a demagogue.

The latent momentum of an externalized mental model interacts with the active momentum carried by the flow, much as the potential of the field created by a magnet changes the momentum of a nearby paper clip. A charging mob contains more externalized momentum than a quiet one. A shopping mall contains more externalized momentum than a library.

A social group can exhibit a range of behavior depending on the field it is in. An army marching in a parade is a highly orderly, legible and meaningful flow. An army at work in the battlefield is, to the untrained external eye, highly disorderly and illegible. The former is an impoverished expression of its capabilities, meant to simultaneously showcase power and signal that it is under control, while the latter is a full expression of its power.

The army example highlights an important point about social realities. For a given individual, the environmental field (the parade ground or battlefield) combines with the other group members in the field at the same time, to create a *social field.* When there is little momentum or meaning in the primary environmental field, the dynamics of the social field start to dominate.

An environmental field that is deliberately designed to catalyze complex social behaviors generally contains one or more *social objects,* a term coined by cartoonist Hugh MacLeod.

Chairs and tables are among the simplest social objects. Consider the problem of arranging chairs and tables at a large meeting or event.

A meeting venue with semi-circular rows of stadium-style seating will create a more intimate atmosphere and invite participation in what the speaker, at the focal point, has to say. Horizontal rows of chairs on a flat surface, facing a raised stage, will create more distance and dampen participation. Banquet style seating, with groups of six to eight chairs arranged around circular dinner tables, will decentralize and localize the social fields and flows, making it harder to command attention from a central podium. Awards ceremonies organized as banquets have a very different vibe from those organized as a spectacle, with all seats facing a podium.

The analogy to magnetism is a useful one. A field turns natural behaviors – randomly oriented paper clips for instance – into behavior that exhibits some order. Through this creation of order out of chaos, our natural behaviors can become effective procedural behaviors.

6.2.1   Orchestrating Field-Flow Complexes

In our discussion of basic decision patterns, we noted that the fourth kind, procedural patterns, depended on externalized systems and processes. Systems and processes are the simplest kind of field-flow complex. They are to social reality what calculative rationality is to narrative rationality.

A set of systems and processes, or a system-process complex, is usually the codified embedding of the mental models of a few centralized decision-makers, working according to highly platonic logic, and driven by a great desire for legibility and meaning from a central, but external, point of view. Just as calculative rationality often fails due to its impoverished view of planning and decision-making, system-process complexes often fail because they prioritize legibility and order over effectiveness.

The governing aesthetic of system-process thinking is what Scott[43] calls "authoritarian high modernism." When system-process complexes fail, it is often due to the same pathology we noted before: overweening, impoverished design crushing an ongoing, organic dance.

System-process thinking is a specific kind of authoritarian high modernism usually known as Taylorism, after the ideas of Frederick Winslow Taylor. He was a neurotic genius, management pioneer and author of the seminal 1919 book, *The Principle of Scientific Management*.[46] Taylor's influence led to some of the worst authoritarian high-modernist debacles of the twentieth century, and as a result his ideas are commonly demonized today.

It is important to recognize, however, that his ideas and methods, carefully applied, have proved highly valuable. Interval logic, the idea we encountered in Chapter 2, is a refined descendant of the Taylorist idea of Gantt charts. Where Taylorism goes wrong is when it exceeds its bounds of applicability (and admittedly, Taylor's presumption that his broader philosophy was objectively "scientific" in some sense was a major contributing factor).

Taylorism is commonly criticized for its dehumanizing effects, but for us, the more important problem is that it is also ineffective with respect to many of the problems it attempts to solve.

To understand the limits of Taylorist system-process thinking, it is only necessary to imagine applying its methods (such as time and motion studies) to a problem such as "how can I throw a great dance party?"

For a simpler illustration of the limits, recall the problem of restless elevator riders impatiently pushing the call button too many times. A typical Taylorist thinker would add a feedback signal, such as a display that shows the current location of all elevators, or a sign that says "please do not repeatedly push the button."

To arrive at artistic solutions like putting a mirror across from the bank of elevators requires moving beyond the systems-process mindset.

Artistic approaches to field-flow complexes exploit the natural chemistry among mental models in richer ways. The key to such approaches is a tasteful blending of preparatory design and participatory dance. Successful field-flow complexes can neither be designed nor danced into existence. They must be *orchestrated*. A conductor or choreographer, rather than a pure designer or dancer, is necessary.

To inform and catalyze such artistry through design and dance, the choreographer of a field-flow complex must have a sense of all the internal and embedded mental models involved in a situation, along with their mutual levels of legibility and meaningfulness. A wedding planner may choose centerpieces ahead of time (design over dance), while a hostess at a party might instinctively circulate to drive up the energy of a party (dance over design). In either case, at the heart of effective orchestration is an intuitive grasp of the interaction of mental models.

The weakest interactions are at the level of tempo, while the strongest involve the construction of shared meaning.

To get two economists to waltz together, all you need is the right kind of music, enough alcohol and a savvy host or hostess. To write an award-winning paper together, the two economists must share a language and have complementary doctrines and intellectual strengths. They will require, as an enabling social context, the complex social fields and flows of an entire academic community.

But even in this complex case, tempo matters: scholarly collaborators, like dancers at a party, must have complementary enactment styles (an idea often oversimplified to "on the same wavelength"). If their enactment styles do not create the right mix of harmony and creative dissonance, they will not be able to work together. Something as trivial as different preferred talking speeds, or a clash of interruption norms (Chapter 2) can get in the way of collaboration.

It is tempting to further formalize the ideas and concepts we have talked about into some sort of codified artistic process. To avoid falling victim to the very high-modernist disease we are attempting to cure, it is important to remain focused on a rich set of examples, rather than the ideas themselves.

Sensitizing yourself to legibility, meaning, ongoing processes of externalization and internalization, latent momentum and tempo is a matter of practice. The skill I introduced in Chapter 1 is really the foundation for the more refined and cultivated ways of seeing we are talking about here.

Ponder these examples, most of which we've encountered previously. What are the relevant mental models in play? Are they internalized or externalized? What is the level of momentum in the situation? What are the levels of mutual legibility and shared meaning? Can you tease apart the pure tempo-level interactions from interactions where shared meaning is being constructed? Can you identify key fields and flows?

- A kitchen in a busy restaurant (Chapter 1)

- A business meeting (Chapter 1)

- You, cooking in your kitchen (Chapter 1)

- A driver merging onto a highway (Chapter 2)

- An Arab-Israeli meeting (Chapter 2)

- A courtroom, with a trial in progress (Chapter 3)

- A wedding (Chapter 4)

- A large dinner party (Chapter 3)

- An assembly line (this chapter)

- Robinson Crusoe on an island

- A hostage negotiation

- A busy store (see Paco Underhill's classic, *Why We Buy*[47] for insight into this example)

Each of these examples involves complex interactions among mental models and environments. Groups cannot be realistically studied unless you take into account the effects of the field-flow complexes that they inhabit. This is a natural consequence of narrative rationality: you can only understand a group with reference to the specific narrative within which it is situated.

The environment, manifesting as it does the externalized mental models of individuals who may or may not be present, is an active participant in any real-world group situation. The designers of the chairs and tables, the providers of coffee and cookies, and the interior designer who has laid out the room and its lighting, all participate in every group situation that plays out in a room. Any business conducted during a game of golf is influenced by the designers of the course and the clubs used by the players, *in absentia*. A negotiation on a golf course is very different from one in a board room.

6.2.2 Shared Mental Models

As field-flow complexes get larger and richer in their behavior, even the most artistic individual cannot orchestrate them single-handedly.

As the scale of human activity increases, processes of deliberate design and centralized orchestration by a few give way to emergent dynamics and decentralized orchestration by many. Dance begins to overwhelm design, and at the largest scales such as the global economy, effectively replaces it altogether. Through several millennia of technological development, we have gradually been able to introduce deliberate design at larger and larger scales, but as the examples in *Seeing Like a State* illustrate, at the largest scales, deliberate design remains effectively impossible. This is the central reality which must be faced by all organizations.

So how *do* large and coherent organizations emerge at all?

They emerge as an outcome of growth and evolution processes, rather than construction processes, and as such cannot be comprehended *a pri-*

ori by a human designer. They only make sense with the benefit of hindsight, and even then, only make partial sense. Organizations that are grown are fundamentally less legible than those that are constructed.

For these processes to be successful, the mental models that seed any generative process of growth must have sufficient levels of mutual legibility, and harmonize effectively at the level of tempo. In other words, they must be *shared* mental models, capable of sustaining a *shared* situation awareness of relevant realities.

Shared mental models must include organizational self-archetypes and doctrines, just as individual mental models must include individual-level self-archetypes and doctrines.

For organizations that exist within a larger ecosystem that includes other organizations, such as the set of businesses competing in a market, organizational mental models must also include archetypes for other organizations. So a particular business might, for instance, view itself as a creative, go-getter organization, and its older competitors as staid and slow.

Where complex realities are to be grown rather than constructed, the role of the individual orchestrator is limited to catalyzing the emergence of the right shared mental models in the early stages. This includes planting the seed of an organizational self-archetype and doctrine that is appropriate to the *raison d'etre* of the organization. Once the process is underway, the orchestrators can do little: the organization dances itself into existence and self-awareness.

Organizational archetypes and doctrines rely on metaphors to anchor their overarching sense of self.

Gareth Morgan, in *Images of Organization*,[48] showed that we commonly think about organizations through one of eight metaphors. The eight metaphors are: organizations as machines, as brains, as organisms, as cultures, as political systems, as psychic prisons, as systems of change and flux, and as systems of domination.

The interesting thing is that all the metaphors, except for the Taylorist mechanistic one, naturally comprehend growth and evolution. Where change in the machine metaphor is a process of stepwise re-engineering, in the other, more organic metaphors, change is a process of generative growth, ontogeny and self-organization.[§]

[§] Such thinking has spread well beyond organizational theory to fields such as software design. The principle of "paving the cowpaths" in user-interface design is based on the

6.3   Exploring Organizations

In the previous sections, I have attempted to sketch out a basic approach that allows us to apply the individual-centric ideas in previous chapters to social reality.

This leads naturally to a study of organizations, at all scales ranging from married couples to civilizations.

This is a vast topic, and one that I frequently write about on my blog. While a complete treatment is beyond the scope of this book, the ideas we have covered so far should allow you to productively ponder key questions such as the following:

1. How do flows stabilize into repeated behaviors and group procedural tactics?

2. When intelligence and mental models are externalized into the environment, do organizations get smarter?

3. As flows stabilize and externalizations are completed, how does ritualization emerge? What creates organizational inertia?

4. Does ritualization make the participating humans stupider than their environments?

5. Do missions and visions represent real thinking, or ritualized and ineffective processes that fail to create shared mental models?

6. Are evolving self-archetypes the same as organizational brand narratives?

7. Do common individual doctrines, such as those in Chapter 3, also correspond to common organizational doctrines?

8. Do organizations, like individuals, experience deep stories?

9. What do the different epochs map to? What constitutes a *cheap trick* or *valley* for an organization?

10. How does an organization make sense of its broader environment?

---

following organizational design metaphor: When designing a network of paved paths through a garden, you should allow people to walk where they will for a few months. Once they've worn out some paths, these should be paved. This avoids the risk of users straying from pre-conceived paths and ruining the turf.

11. Does each organization operate by its own unique narrative rationality, or are there only a few distinct types?

12. How does the idea of an organization's "Grand Narrative" relate to the concept of the Freytag staircase?

13. How can you tell whether a ritualized flow, based on externalized mental models of forgotten founders, is obsolete, or effective but illegible?

14. How do fields and flows decay, degrade and die? How do new ones take root in old organizations?

15. How do organizations die? Must they necessarily die?

16. Do individual-level universal tactics suffice to describe and explain organizational behavior?

While useful first-order answers to most of these questions require only an imaginative reinterpretation of the corresponding individual-level ideas, organizations do differ from individuals in a few key ways.

First, while growth for humans is a largely automatic process that creates adults out of single fertilized eggs, organizations, as we have noted, are orchestrated into existence through a mixture of deliberate design and dance. A comprehensive tempo-based theory of organizations must therefore also consider design-and-dance doctrines in addition to the decision-making and momentum-management doctrines we have considered. Each such doctrine must codify design principles that relate to the management of fields and flows, in relation to specific domains.

Examples of such doctrines include authoritarian high modernism, which we have already discussed, and its antithesis, orchestration (which I like to call *laissez-faire* low naturalism). Time-honored aesthetic-spiritual practices such as the Japanese *wabi-sabi* and the Chinese *feng-shui* can also be interpreted as design-and-dance doctrines. More recently, the field of behavioral economics has given rise to a young doctrine that calls itself "libertarian paternalism."[49]

Second, organizations, unlike individuals, can overlap and subsume each other in more intimate ways than individuals can. Individuals can think in terms of figurative mind meld with each other, and feel either a strong or weak sense of belonging to organizations. Among themselves, though, organizations can interact in deeper ways and exhibit phenomena such as true mergers and acquisitions. This is a simple consequence of the fact that organizations at all scales comprise people.

But people themselves are made of cells, organized in a a set of interwoven organ systems and material flows that dance to chronobiological rhythms. Beyond the limited processes of sexual reproduction and transplant surgeries, human bodies cannot truly interact in rich ways. The metaphoric mapping between organizations and individual humans, at the level of both mind and body, only goes so far. Physics and biology intervene, and you must be careful not to take the metaphor too far in careless ways.

6.4 Skill: Reading Time Cultures

Reflect for a moment on the audacity of what we've attempted to do in this chapter: develop an impressionistic, integrated portrait of at least a dozen complex fields, each of which deserves at least a shelf-load of books. We've touched on organization theory, architecture, urban planning, group dynamics, language and communication, party planning, biology, military science, and the organization of desks. We've ranged from the trivial, such as potted plants and coffee cups, to the sublime, such as the metaphysics of dance.

As you reflect, you will certainly notice flaws and problems. That is inevitable when you attempt to see the world in a grain of sand. But what is remarkable is that the idea of tempo is able to sustain even a moderately coherent account of so much territory.

The reason tempo has proved to be such a broadly useful notion is that time is so fundamental to the universe. It is, as statisticians like to say, a fertile variable. It shows up everywhere.

So to conclude this chapter, I'd like to outline the basics of a rather philosophical skill that is nevertheless very useful: reading time cultures. It is an abstract analog to the skill we started with, tempo doodling.

To read the time culture of any organization, ranging from a two-person marriage to all of human civilization (or for that matter, a non-human organization like a pride of lions in the savannah), try the following:

1. Read, observe and participate in your target culture as much as possible; immerse yourself in it

2. Note the important rhythms, basal energy levels, and emotional temper of the culture

3. Try to experience the core event stream of the culture, using the descriptive language of temporal interval logic we introduced in Chapter 2

4. Try to experience the culture's sense of historicity and entropy; its views on progress and decline, as embodied in its codified rituals

5. Learn how the culture frames risk, knowledge and the unknown

6. Sketch the Freytag staircase for the culture as a whole, from its mythological beginnings to its prophesied end, keeping in mind that you may find something that is not a staircase at all, but an eternal cyclic narrative

This outline of how to read a time culture is based on, and beautifully illustrated by, Clifford Geertz' classic study of Bali, *Deep Play: Notes on the Balinese Cockfight*.[50] Reading a time culture this way is a distinctly political act, since you cannot hope to suspend your native narrative time entirely. To get a sense of the fascinating politics of time, you may want to read Jeremy Rifkin's entertaining polemic, *Time Wars*,[51] and Jay Griffiths more recent *A Sideways Look at Time*.[52]

Unlike its cousin space, time is a very intimate dimension. The entire world can share an instant of time in a way that it cannot share a location in space. There is a reason we remember 9/11 first as a date and only then as a place. Rhythms in time bind us more closely than patterns in space.

In Chapter 4, we introduced a notion of narrative time: a textured, situation-dependent time characterized by its periodic, aperiodic, and entropic (directional) aspects. We distinguished this from the sort of featureless, smooth and reversible platonic time that anchors calculative rationality. There is a reason anthropologists of time rail against atomic clocks. Atomic clocks are very *precise*, but they are not particularly *accurate* for all purposes. You can measure the time *when the cows come home* far more accurately by waiting to hear a moo, than by watching an atomic-clock display. As an intermediary between the sun and the rooster, the atomic clock is fairly useless.

While atomic-clock time is no more unnatural than Nile-flooding time (cesium atoms, after all, are as natural as anything else), it suffers from its impoverished, high-legibility design. While this is useful for orchestrating the motions of communication satellites and managing Internet traffic, in other contexts it can smooth the texture of local narrative time to the point that it turns bland.

Robert Levine's *A Geography of Time*,[8] which I mentioned in passing in Chapter 2, is as valuable to narrative-time tourists as a *Lonely Planet* guide is to those who limit their travels to space.

So when you next travel to an island destination like Hawaii, you need to reset two clocks. When you reset your wristwatch to the local time, remember also to reset your narrative time clockless clock to *island time*. The one will help you catch your flight back home, but the other will allow you to actually arrive at your destination in a narrative sense.

Chapter 7

Conclusion

Our lives are suffused by a deep tension between contemplation and action. Contemplation without action – the spectator's mode of life – seems somehow empty and devoid of meaning. Action without contemplation on the other hand, seems a base way of life. We looked for meaning, and skeptically questioned meaning, in everything from cooking to Shakespeare.

Of the many conceits with which we humans burden ourselves, perhaps none is deeper than the conceit that our lives, unlike those of other animals, must be meaningful. And so we conclude that the unexamined life is not worth living. Equally, most of us conclude, the unlived life is not worth examining.

If there is an overarching theme to this book, it is ultimately this tension between action and contemplation. A tension that causes us to swing between a greedy, grasping engagement of life, and a tentative, doubtful withdrawal from it. A mode of life that doubts the possibility of meaning sufficiently to choose action, and believes in the possibility sufficiently to be tempted into reflection.

It is my hope that the ideas in this book will help you turn this tension into a source of creative-destructive energy to fuel your life.

7.1 Additional Resources

An evolving website and email newsletter associated with this book can be found at **http://tempobook.com**. I will gradually be adding supplementary material, learning resources and multimedia content to the website.

If the broader themes in this book interest you, I invite you to join me and the regular readers of my blog at:

http://ribbonfarm.com

If you have feedback, requests or suggestions for future editions of the book, please do get in touch at **vgr@ribbonfarm.com**.

Acknowledgments

Any work of broad synthesis and integration must rely particularly heavily on the work of others, and this book is no exception. The list of people who have helped shaped this book is a long one.

My doctoral advisor, Pierre T. Kabamba at the University of Michigan, and my postdoctoral advisor, Raffaello D'Andrea at Cornell (now at ETH, Zurich), helped me establish firm intellectual foundations in the discipline (systems and control theory) that has most shaped my thinking, and indulged and encouraged my tendency to wander very far afield from my home base. Only now, years later, do I fully appreciate how much I benefitted from those years of unfettered exploration.

Also at Michigan, Ed Durfee, Demosthenis Teneketzis and Martha Pollack helped me broaden my intellectual horizons beyond the narrow confines of my home discipline, to domains such as artificial intelligence, linguistics, operations research, economics, and game theory.

At Xerox, a few encounters proved crucial in the development of my thinking about organizations and interpersonal dynamics. Steve Hoover (currently the CEO of PARC), has forgotten more about organizational behavior, effective management and interpersonal dynamics than I will ever know. To him I owe my knowledge and understanding of many subtle lessons in the practice of management.

Among the many colleagues with whom I've had the pleasure of working, the two co-conspirators with whom I worked most closely at Xerox, Dave Vandervort and Jesse Silverstein, taught me more about decision-making than they perhaps realize, not merely by modeling effective decision-making styles very different from my own, but by keeping me honest and accountable for my own decision-making style.

Two good friends, Michael Allers and Max Montesino, introduced me to subjects that have been particularly crucial for this book: the

153

philosophy of language, and the theory of conceptual metaphor. These subjects are so far from my usual stomping grounds in engineering that this book would literally not have been possible, had I not met them. These serendipitous intellectual encounters were catalyzed by the Telluride House at the University of Michigan. It was also at Telluride House that I met Adam Hogan, another good friend, who designed the cover for this book and helped me figure out many of the intricacies of self-publishing.

ESD generously helped out at the last minute with a round of copy-editing.

Three writers have been extremely generous with their time, counsel and encouragement over the last few years, as I attempted to learn the craft that they have mastered: David Allen, author of *Getting Things Done*; Daniel Pink, author of *Free Agent Nation, A Whole New Mind* and *Drive*; and Erik Marcus, author of *Meat Market*.

Closer to home, my wife Mee Yong has endured several years of my obsessive absorption in reading and writing, as I attempted to balance the competing demands of a full-time job, a growing blog and this book project. All three activities grew increasingly demanding over the last several years, and I am guilty of neglecting her and the rest of my family more than I like to admit.

Two virtual institutions that have been made this book possible are the open source communities behind the WordPress blogging platform and LATEX, the typesetting program I used to create this book.

And finally, I must thank the readers of my blog. Their apparently inexhaustible capacity for absorbing and improving upon anything I write is ultimately what has allowed me to develop my voice as a writer. Three readers in particular, Davison Avery, Sean Murphy and Ho-Sheng Hsiao, provided a great deal of thoughtful feedback and discussion during the development of this book.

Bibliography

[1] Fredrick P. Brooks. *The Mythical Man-Month*. Addison-Wesley; 2nd edition, 1995.

[2] Tom Vanderbilt. *Traffic: Why we drive the way we do (and what it says about us)*. Random House, 2008.

[3] Marvin Minsky. *The Society of Mind*. Simon and Schuster, 1988.

[4] Bill Tancer. *Click*. Hyperion, 2008.

[5] Carl von Clausewitz. *On War*. Brownstone, 2009.

[6] Alfred Thayer Mahan. *The Influence of Sea Power Upon History, 1660-1783*. Dover, 1987.

[7] Robert Coram. *Boyd: The Fighter Pilot who Changed the Course of War*. Back Bay Books, 2004.

[8] Robert V. Levine. *A Geography of Time*. Basic Books, 1998.

[9] Dean Ornstein. *On the Experience of Time*. blah, 1933.

[10] Philip Zimbardo and John Boyd. *The Paradox of Time*. blah, 2009.

[11] Mica R. Endsley, Betty Bolte, and Debra G. Jones. *Designing for Situation Awareness*. CRC Press, 2003.

[12] Daniel H. Pink. *Free Agent Nation*. Business Plus, 2002.

[13] William Rathje. *Rubbish: The Archeology of Garbage*. University of Arizona Press, 2001.

[14] Charles Fine. *Clockspeed*. Basic Books, 1999.

[15] Keith Johnstone. *Impro: Improvisation and the Theater*. Routledge, 1987.

[16] Michael Bratman. *Intention, Plans and Practical Reason*. Harvard, 1987.

[17] Isaih Berlin. *The Hedgehog and the Fox: An Essay on Tolstoi's View of History*. Ivan. R. Dee, 1993.

[18] Eric Berne. *Games People Play*. Ballantine Books, 1997.

[19] Kenneth Grahame. *The Wind in the Willows*. Palazzo, 2008.

[20] Philip Pullman. *His Dark Materials Trilogy*. Laurel Leaf, 2003.

[21] Jennifer von Bergen. *Archetypes for Writers*. Michael Wiese Productions, 2007.

[22] C. Rajagopalachari. *The Mahabharata*. Bharatiya Vidya Bhavan, 1951.

[23] Nicholas Nassim Taleb. *The Black Swan*. Random House, 2nd edition, 2010.

[24] William Duggan. *Strategic Intuition*. Columbia University Press, 2007.

[25] Gustav Freytag. *Freytag's Technique of the Drama*. Bibliobazaar, 2008.

[26] Joseph Campbell. *The 1000 Faces of the Hero*. New World Library, 3rd edition, 2008.

[27] Alison Gopnik. *The Philosophical Baby*. Farrar, Straus and Giroux, 2009.

[28] Michael R. Garey and David S. Johnson. *Computers and Intractability*. W. H. Freeman, 1979.

[29] Stephen A. Cook and David G. Mitchell. Finding hard instances of the satisfiability problem. In *Proc. DIMACS workshop on Satisfiability Problems*, 1997.

[30] Peter Cheeseman, Bob Kanefsky, and William M. Taylor. Where the really hard problems are. In *Proc. IJCAI-91*, pages 163–169, Sydney, Australia, 1991.

[31] Gregory J. Chaitin. *Meta Math*. Vintage, 2006.

[32] Dan P. McAdams. *The Redemptive Self: Stories Americans Live By*. Oxford University Press, 2005.

[33] Albert Camus. *The Myth of Sisyphus*. Vintage, 1991.

[34] Jennifer Ackermann. *Sex Sleep Eat Drink Dream*. Mariner Books, 2008.

[35] Russell G. Foster and Leon Kreitzman. *Rhythms of Life*. Yale University Press, 2005.

[36] Stuart A. Kauffman. *Investigations*. Oxford University Press, 2002.

[37] George Lakoff and Mark Johnson. *Metaphors we Live By*. University of Chicago Press, 1980.

[38] Rachel Kaplan and Stephen Kaplan. *The Experience of Nature: A Psychological Perspective*. Cambridge University Press, 1989.

[39] Gary Klein. *Sources of Power*. MIT Press, 1998.

[40] Malcolm Gladwell. *Blink*. Back Bay Books, 2007.

[41] Daniel C. Dennett. *Consciousness Explained*. Back Bay Books, 1992.

[42] Oliver Sacks. *The Man Who Mistook His Wife for a Hat*. Touchstone, 1998.

[43] Jamces C. Scott. *Seeing Like a State*. Yale University Press, 1999.

[44] David Allen. *Making it All Work*. Penguin, 2009.

[45] John R. Searle. *The Construction of Social Reality*. Free Press, 1997.

[46] Frederick W. Taylor. *The Principles of Scientific Management*. Dover, 1998.

[47] Paco Underhill. *Why We Buy*. Simon and Schuster, 2000.

[48] Gareth Morgan. *Images of Organization*. Sage Publications, 1996.

[49] Richard H. Taylor and Cass R. Sunstein. *Nudge*. Yale University Press, 2008.

[50] Clifford Geertz. *The Interpretation of Cultures*. Basic Books, 1977.

[51] Jeremy Rifkin. *Time Wars*. Simon and Schuster, 1989.

[52] Jay Griffiths. *A Sideways Look at Time*. Tarcher, 2004.

Lightning Source UK Ltd.
Milton Keynes UK
UKOW05f0620300617

304414UK00001B/259/P